Enhancing Space Resilience Through Non-Materiel Means

Gary McLeod, George Nacouzi, Paul Dreyer, Mel Eisman, Myron Hura,
Krista S. Langeland, David Manheim, Geoffrey Torrington

RAND Project AIR FORCE

Prepared for the United States Air Force

Approved for public release; distribution unlimited

For more information on this publication, visit www.rand.org/t/RR1067

Library of Congress Cataloging-in-Publication Data is available for this publication.

ISBN: 978-0-8330-9313-4

Published by the RAND Corporation, Santa Monica, Calif.

© Copyright 2016 RAND Corporation

RAND® is a registered trademark.

Limited Print and Electronic Distribution Rights

This document and trademark(s) contained herein are protected by law. This representation of RAND intellectual property is provided for noncommercial use only. Unauthorized posting of this publication online is prohibited. Permission is given to duplicate this document for personal use only, as long as it is unaltered and complete. Permission is required from RAND to reproduce, or reuse in another form, any of its research documents for commercial use. For information on reprint and linking permissions, please visit www.rand.org/pubs/permissions.html.

The RAND Corporation is a research organization that develops solutions to public policy challenges to help make communities throughout the world safer and more secure, healthier and more prosperous. RAND is nonprofit, nonpartisan, and committed to the public interest.

RAND's publications do not necessarily reflect the opinions of its research clients and sponsors.

Support RAND

Make a tax-deductible charitable contribution at
www.rand.org/giving/contribute

www.rand.org

Preface

Current U.S. national security space (NSS) architectures were developed assuming that space was a sanctuary. However, as stated in the *National Security Space Strategy*, the space environment "is becoming increasingly congested, contested, and competitive."[1] In recent analyses of alternatives for future NSS systems, the U.S. Department of Defense (DoD) and the U.S. Air Force have begun exploring how the space segment can become more resilient to potential adversary actions, as well as to system failures and the harsh environment of space. However, enhancing the resilience of NSS capabilities must occur in today's financially constrained environment, and changes to the space segment will likely be expensive.

To provide a complete look at resilience and possibly realize some benefit in the near term and at lower cost, the Air Force has tasked RAND to identify non-materiel means—doctrine, organization, training, materiel, leadership and education, personnel, facilities, and policy (DOTMLPF-P)[2]—to enhance space resilience. Thus, this report should be of interest to those in the U.S. government space community looking to enhance the resilience of current and future U.S. space systems. Readers are also likely to be interested in the project's three companion reports:

- Myron Hura, Gary McLeod, and George J. Nacouzi, *Enhancing Space Resilience Through Non-Materiel Means: Appendix B—Missile Warning Mission Case Study*, Santa Monica, Calif.: RAND Corporation, 2016, not available to the general public.
- Krista S. Langeland, David Manheim, Gary McLeod, and George J. Nacouzi, *How Civil Institutions Build Resilience: Organizational Practices Derived from Academic Literature and Case Studies*, Santa Monica, Calif.: RAND Corporation, RR-1246-AF, 2016.
- Paul Dreyer, Krista S. Langeland, David Manheim, Gary McLeod, and George J. Nacouzi, *RAPAPORT (Resilience Assessment Process and Portfolio Option Reporting Tool): Background and Method*, Santa Monica, Calif.: RAND Corporation, RR-1169-AF, 2016.

This research was sponsored by the commander, Air Force Space Command, and was conducted within the Force Modernization and Employment Program of RAND Project AIR FORCE as part of a fiscal year 2014 project, "Space Resilience: Developing a Strategy for Balancing Capability and Affordability with Resilience." The information presented here is current as of September 2014.

---

[1] U.S. Department of Defense and Office of the Director of National Intelligence, *National Security Space Strategy*, Unclassified Summary, Washington, D.C., January 2011, p. 1.

[2] DOTMLPF-P is a DoD acronym for the eight possible non-materiel elements involved in solving warfighting capability gaps: doctrine, organization, training, materiel, leadership and education, personnel, facilities, and policy. The "materiel" category does not include new defense acquisition programs (such as a new space segment); however, it does include some ground segment materiel needs, such as facilities and information technology. A lowercase *m* is often used in the acronym: DOTmLPF-P.

RAND Project AIR FORCE

RAND Project AIR FORCE (PAF), a division of the RAND Corporation, is the U.S. Air Force's federally funded research and development center for studies and analyses. PAF provides the Air Force with independent analyses of policy alternatives affecting the development, employment, combat readiness, and support of current and future air, space, and cyber forces. Research is conducted in four programs: Force Modernization and Employment; Manpower, Personnel, and Training; Resource Management; and Strategy and Doctrine. The research reported here was prepared under contract FA7014-06-C-0001.

Additional information about PAF is available on our website:
http://www.rand.org/paf/

This report documents work originally shared with the U.S. Air Force on September 23, 2014. The draft report, issued in December 2014, was reviewed by formal peer reviewers and U.S. Air Force subject-matter experts.

Table of Contents

Figures

Tables

Summary

The U.S. Department of Defense (DoD) is now treating the space domain as a congested, contested, and competitive environment. The Air Force has also realized that the national security space (NSS) enterprise must undergo a transformation to help it survive in this new environment and has been looking at ways to make the enterprise more resilient. Consequently, Air Force Space Command (AFSPC) has been considering options in which the future NSS architecture continues to provide the needed mission capabilities in the face of hostile activities. Novel options that have been considered include hosted payloads, disaggregated architectures, new orbits, and international cooperation, as well as various other creative concepts.

However, most of the efforts have been focused on the space segment (i.e., the on-orbit assets). Other important and effective options reside in the non-materiel contributions of the space enterprise: doctrine, organization, training, materiel, leadership and education, personnel, facilities, and policy (DOTMLPF-P).[3] Some of the elements of DOTMLPF-P are direct enablers of space services (e.g., facilities and personnel), and others have an indirect—but just as critical—role (e.g., policy and training). Additionally, DOTMLPF-P improvements can provide improvements to the resilience of the space mission independent of the space segment[4] and, in some cases, at a relatively low cost, which is important in the budget-constrained environment faced by DoD.

Objective

We were tasked by AFSPC to assess and modify some of the elements of DOTMLPF-P to improve space mission resilience. The objective was to identify and recommend near-term (i.e., "low-hanging fruit"), mid-term, and far-term DOTMLPF-P actions that will help improve space mission resilience, independent of the space segment architecture. The recommendations would consider ease of implementation and cost to implement, as well as overall improvements to the space mission resilience.

Space Resilience

DoD has defined *resilience* in the context of space as

---

[3] DOTMLPF-P is a DoD acronym for the eight possible non-materiel elements involved in solving warfighting capability gaps: doctrine, organization, training, materiel, leadership and education, personnel, facilities, and policy. The "materiel" category does not include new defense acquisition programs (such as a new space segment); however, it does include some ground segment materiel needs, such as facilities and information technology. A lowercase *m* is often used in the acronym: DOTmLPF-P.

[4] However, without changes to the space segment, improving an operator's abilities to respond to a contested environment will provide an incremental improvement in the resilience of the space mission. Clearly, survivability needs to be designed into the space segment, not left as an afterthought to be handled by the ground element.

[t]he ability of an architecture to support the functions necessary for mission success with higher probability, shorter periods of reduced capability, and across a wider range of scenarios, conditions, and threats, in spite of hostile action or adverse conditions.[5]

There are a number of approaches being discussed to increase the resilience of a space architecture. They include disaggregation, distribution, diversification, protection, proliferation, and deception.[6]

At the ground segment, however, enhancing resilience really devolves to space protection. The other five categories, as usually defined, apply to how one has already implemented the space and ground segments. Space protection, better described as self-protection for our purposes, involves passive and active measures that the satellite can take to defend itself—often with the assistance of an operator—assuming that such capabilities already are resident on board the satellite. Thus, while we use the term *resilience* throughout the report, when we discuss specific recommendations for the Air Force, they are primarily in the context of space protection.

Approach and Scope

During the execution of this project, we conducted a comprehensive literature review that looked at how different civil industries (e.g., hospitals) address resilience, with the expectation that our findings would have applicability to the Air Force space community. We met with a number of space organizations[7] to better understand how the current DOTMLPF-P elements are affecting the overall space mission resilience. We heard what the current conditions are, what some are doing to improve resilience, and suggestions on how to further improve it. We collected these findings, assessed and refined many of the suggestions provided, and added a number of our own options for improvement. We then developed an integrated methodology, based on a space protection operational concept, to help assess the contributions of these non-materiel options to resilience.

While the results of our research may apply to other space missions, we focused on force enhancement missions[8] because these capabilities are likely to be targeted by a potential adversary who wants to degrade the warfighting effectiveness and efficiency of our air, ground, and naval forces. We also focused on the tactical or squadron level (i.e., space operations squadrons [SOPSs] and space warning squadrons [SWSs]), where command and control of Air

---

[5] DoD, "Space Policy," DoD Directive 3100.10, October 18, 2012.

[6] During preparation of the final report, the authors learned that the Air Force, following DoD's recent Strategic Portfolio Review for Space, is leaning away from disaggregation, a concept advocated in an AFSPC white paper released in August 2013, toward a broader theory informally known as *space protection*, and that disaggregation is now seen as just one element of a broader space-protection solution (Mike Gruss, "Disaggregation Giving Way to Broader Space Protection Strategy," *Space News*, website, April 26, 2015).

[7] *Organizations* here includes those at the squadron to headquarters level and includes both Air Force and non–Air Force organizations (e.g., NASA, National Oceanic and Atmospheric Administration [NOAA], and a commercial provider).

[8] Force enhancement missions include positioning, navigation, and timing; military satellite communications; missile warning; environmental monitoring; and intelligence, surveillance, and reconnaissance.

Force satellites occurs, and on the operational level (i.e., the Joint Space Operations Center [JSpOC], which is responsible for operational employment of worldwide joint space forces).

Overarching Recommendations

Resilience as a Priority

One issue that was brought up by all the space operators with whom we met is the apparent lack of "demonstrated" priority to resilience by the leadership. Although they are aware that the commander, AFSPC, and other senior Air Force and DoD leaders from the space community discuss its importance,[9] the fact that resilience is a priority has not promulgated formally to space squadrons in the form of detailed implementation actions. Changing the prevalent mindset/culture that "space is a sanctuary" within the rank and file of the space operator community will require that

- Space leadership define priorities and provide resources for non-materiel space resilience activities.

We expect that developing and implementing some of the recommendations provided in this report will help in both improving the resilience of the space enterprise and changing the mindset of the involved personnel.

Space Protection Concept of Operations

Because of the interdependence of the various DOTMLPF-P elements, making a few changes will not result in the desired improvement. We developed a set of implementation options to improve resilience based on a notional space protection operational concept—namely, enhancing the capability of space operators to respond, in a timely and effective manner, to adversary counterspace actions. To do so, operators need actionable information, appropriate organization and tactics, and dynamic command and control, supported by appropriate tools and decision aids, relevant training and exercises, and qualified personnel brought into the career field. While this operational concept is a good starting point, we recommend that

- AFSPC develop a formal, end-to-end, space protection concept of operations (CONOPS) that captures all the elements needed to improve resilience.

In addition, when developing the CONOPS, it may be time for the space community to relax its centralized control and centralized execution in certain situations, such as responding to adversary counterspace actions, and follow the tenet of centralized control and decentralized execution, which is considered crucial to the effective application of airpower.

---

[9] The fact that U.S. satellites are at risk was made very clear to the nation on April 26, 2015, when CBS aired a segment called "The Battle Above" on *60 Minutes*, in which senior Air Force leadership described the contested space environment (CBS, "The Battle Above," *60 Minutes*, April 26, 2015).

Detailed Recommendations

As discussed in Chapter 4 and summarized in Chapter 6, we developed a detailed set of non-materiel implementation options to improve resilience based on our notional space protection operational concept. We grouped the recommendations for either near-term (less than one year) or far-term implementation (three to six years); they are summarized below.[10] First we list the general mitigation to an identified shortfall and then the specific implementation option for that time frame.

Near-Term Recommendations

- Determine and implement the best means by which the JSpOC Intelligence, Surveillance, and Reconnaissance Division (ISRD) can provide timely counterspace threat advisories and indications and warning (I&W) to Wing intelligence (INTEL) and SOPSs/SWSs: Forward information by chat or email and update website.
- Determine and implement the best means by which the JSpOC Combat Operations Division (COD) can provide timely space weather effects advisories to SOPSs/SWSs: Forward information by chat or email and/or establish website with timely updates.
- Introduce space protection lead at SOPSs/SWSs and JSpOC: Assign space protection as secondary duty to existing crew position.[11]
- Establish process for developing tactics for likely counterspace threats and make their development a priority: SOPSs/SWSs weapons and tactics develop job aids and procedures.
- Review chain of command and determine which command level should have responsibility/authority for various responses to adversary counterspace actions: JSpOC Combat Plans Division develops courses of action for likely adversary threats and establishes rules of engagement that authorize lowest levels of command to provide more timely response.
- Update training process to include recognizing and responding to adversary counterspace actions: SOPSs/SWSs establish on-the-job training for job aids and procedures.

Far-Term Recommendations

- Transfer space order of battle (OB) responsibility from National Air and Space Intelligence Center to JSpOC ISRD: Establish a cadre of government civilians to maintain space OB.
- Determine and implement best means by which JSpOC ISRD can provide timely counterspace threat advisories and I&W to Wing INTEL and SOPSs/SWSs: Define requirements by which JSpOC Mission System (JMS) can be the mechanism for intelligence updates across the space enterprise and phase capability into JMS program.

---

[10] Most of the mid-term implementation options were enablers of far-term options, and their costs were included in the far-term options.

[11] During preparation of the final report, the authors learned that an operator position, a space protection duty officer, has been identified at JSpOC and that additional crew members are currently being trained in response to this need.

- Determine and implement best means by which JSpOC COD can provide timely space weather effects advisories to SOPSs/SWSs: Define requirements for JMS to be the mechanism for providing space weather effects advisories and phase capability into JMS program.
- Introduce space protection lead at SOPSs/SWSs and JSpOC: Create a new crew position.
- Establish process for developing tactics for likely counterspace threats and make their development a priority: SOPSs/SWSs submit tactics to formal tactics, techniques, and procedures process for requisite testing and documentation.
- Review chain of command and determine which command level should have responsibility/authority for various responses to adversary counterspace actions: Define requirements by which JMS can be the mechanism for enabling higher echelons to exercise command by negation and phase capability into JMS program.

Rough Order of Magnitude Costs

For the near-term options listed above, we estimate the rough order of magnitude (ROM) nonrecurring engineering (NRE) cost of implementation to be between $2.5 million and $3.6 million. Similarly, for the far-term options listed above, we estimate the ROM NRE cost to be between $109 million and $166 million, and the ROM recurring (REC) cost to be between $4 million and $5.4 million per year.

Acknowledgments

We would like to thank the project's action officer, Lt Col Steve Lindemuth, chief, Architectures and Support Branch, Air Force Space Command (AFSPC/A5XA), for his assistance during the course of the project, which included arranging meetings with subject matter experts at Headquarters, Air Force Space Command (HQ AFSPC). In addition to those meetings, our research involved discussions with subject matter experts at the following organizations:

Air Force Space Command:
- 50 Operations Group, 1 Space Operations Squadron, 2 Space Operations Squadron, 3 Space Operations Squadron, 4 Space Operations Squadron, and 22 Space Operations Squadron, Schriever Air Force Base (AFB), Colo.
- 460 Operations Group and 2 Space Warning Squadron, Buckley AFB, Colo.
- 11 Space Warning Squadron, Schriever AFB, Colo.
- Advanced Space Operations School, Peterson AFB, Colo.
- Operations Support Division, 614 Air and Space Operations Center, Vandenberg AFB, Calif.

National Reconnaissance Office:
- Survivability Assurance Office, Chantilly, Va.
- Space Operations Squadron, Aerospace Data Facility—Denver, Buckley AFB, Colo.
- Space Operations Squadron, Aerospace Data Facility—East, Ft. Belvoir, Va.
- Intelligence Support Division, National Reconnaissance Operations Center, Chantilly, Va.

Pentagon:
- DoD Executive Agent for Space
- Space and Intelligence Office, Office of the Under Secretary of Defense (Acquisition, Technology, and Logistics)
- Office of the Deputy Assistant Secretary of Defense (Space Policy)
- Deputy Director, Space Operations, Headquarters, U.S. Air Force (AF/A3S)

Other Military:
- Intelligence, Surveillance, and Reconnaissance Division, Joint Space Operations Center, Vandenberg AFB, Calif.
- Space Operations and Integration, U.S. Air Force Warfare Center, Nellis AFB, Nev.
- 533 Training Squadron, Air Education and Training Command, Vandenberg AFB, Calif.
- National Security Space Institute, Air University, Peterson AFB, Colo.

Civil/Commercial Space:
- Space Asset Protection Program Lead, Goddard Space Flight Center, National Aeronautics and Space Administration, Greenbelt, Md.
- Project Lead Engineer, Vehicle Integration Office, Johnson Space Center, National Aeronautics and Space Administration, Houston, Tex.
- Deputy Director, Office of Satellite and Product Operations, National Oceanic and Atmospheric Administration, Suitland, Md.

- Flight Systems Manager, United States Geological Survey, Greenbelt, Md.
- Flight Operations, Intelsat, Washington, D.C.

This work also benefited greatly from extensive discussions on the subject of resilience and space operations with three RAND Air Force Fellows: Col Andrew Kleckner, Col Rose Jourdan, and Col Charles Galbreath. We would also like to thank two RAND colleagues, Yool Kim and Robert Tripp, for their thoughtful reviews. The content and recommendations of this report, however, remain the responsibility of the authors.

Abbreviations

ACC	Air Combat Command
AFB	Air Force Base
AFTTP	Air Force Tactics, Techniques, and Procedures
AFSC	Air Force Special Code
AFSCN	Air Force Satellite Control Network
AFSPC	Air Force Space Command
AoA	analysis of alternatives
AOC	Air and Space Operations Center
ASOpS	Advanced Space Operations School
ASVAB	Armed Services Vocational Aptitude Battery
C2	command and control
CIC	Combined Intelligence Center
COA	course of action
COCOM	combatant command
COD	Combat Operations Division [AOC]
CONOPS	concept of operations
COOP	continuity of operations
CPD	Combat Plans Division [AOC]
DoC	U.S. Department of Commerce
DOI	U.S. Department of the Interior
DOTMLPF-P	doctrine, organization, training, materiel, leadership and education, personnel, facilities, and policy
EUMETSAT	European Organisation for the Exploitation of Meteorological Satellites
FTE	full-time-equivalent
FY	fiscal year
GS	general schedule
HRO	high-reliability organization

I&W	indications and warning
ICT	information and communications technology
INTEL	intelligence
IQT	initial qualification training
ISRD	Intelligence, Surveillance, and Reconnaissance Division [AOC]
JFCC Space	Joint Functional Component Command for Space
JIOC	Joint Intelligence Operations Center
JMS	JSpOC Mission System
JSpOC	Joint Space Operations Center
JWICS	Joint Worldwide Intelligence Communications System
LOE	level of effort
MILPERS	military personnel
MMSOC	Multi-Mission Space Operations Center
NASA	National Aeronautics and Space Administration
NASIC	National Air and Space Intelligence Center
NOAA	National Oceanic and Atmospheric Administration
NRE	nonrecurring engineering
NRO	National Reconnaissance Office
NSOF	NOAA Satellite Operations Facility
NSSI	National Security Space Institute
OB	order of battle
OG	operations group
OJT	on-the-job training
PMEF	Primary Mission Essential Function
REC	recurring
ROE	rules of engagement
ROM	rough order of magnitude
SCRLC	Supply Chain Risk Leadership Council
SIPRNET	Secret Internet Protocol Router Network

SMC	Space and Missile Systems Center
SME	subject matter expert
SNPP	Suomi National Polar-orbiting Partnership
SOPS	space operations squadron
SSA	space situational awareness
SST	Standard Space Trainer
STEM	science, technology, engineering, and mathematics
STRATJIC	U.S. Strategic Command Joint Intelligence Center
SWS	space warning squadron
TRS	training squadron
TTP	tactics, techniques, and procedures
USGS	U.S. Geological Survey
USSTRATCOM	United States Strategic Command
UST	undergraduate space training

1. Introduction

Background

The *National Security Space Strategy* astutely states that "space is becoming increasingly congested, contested, and competitive."[12] Given the critical capabilities that space systems provide to the United States' (and some allies') military operations—in particular, force enhancement—these systems have become an alluring and justifiable military target to potential, and increasingly capable, adversaries. In response, the U.S. Air Force has been reconsidering how to make future space architectures more resilient against potential attacks. Most of these activities are centered around various analyses of alternatives (AoAs) in which a number of approaches to the space segment are being considered, including mission/payload disaggregation, less complex satellites, the use of hosted payloads, and leveraging allied systems.

Non-materiel contributions—doctrine, organization, training, materiel, leadership and education, personnel, facilities, and policy (DOTMLPF-P)—also play a critical role in enabling space support to the warfighter.[13] Any space system architecture needs to consider the supporting infrastructure that enables the mission when assessing resilience. Actually, one can consider modifications to DOTMLPF-P independently from the space segment to improve the overall resilience of the space mission; for example, if operator training is improved, we can expect that the overall mission capability would be incrementally improved. Additionally, most low-cost resilience improvements to the current space mission can only be performed through modifications to DOTMLPF-P, since the only relatively low-cost changes to the on-orbit assets would be limited to software upgrades. Augmenting the space segment, which would involve deploying new satellites, would likely be very costly, and modifying current orbits would likely deplete a significant amount of fuel, thereby reducing the usable life of the satellite.[14]

Note that some organizations prefer to use a lowercase *m* in the acronym to denote non–space segment materiel, since DOTMLPF-P does include some materiel needs, such as facilities and information technology. We also note that even without resilience improvement considerations, some future space architectures will likely require modifications to DOTMLPF-P elements. For example, modifications will be needed to tactics (doctrine), training, and command and control (organization) to support hosted payload operations.

---

[12] U.S. Department of Defense (DoD) and Office of the Director of National Intelligence, *National Security Space Strategy*, Unclassified Summary, Washington, D.C., January 2011, p. 1.

[13] DOTMLPF-P is the DoD acronym that pertains to the eight possible non-materiel elements involved in solving warfighting capability gaps. The *M* in the acronym (which stands for *materiel*) does not include new defense acquisition programs, such as a new space segment. For more information, see Defense Acquisition University, "DOTmLPF-P Analysis," ACQuipedia website, last updated April 15, 2014.

[14] However, without changes to the space segment, improving an operator's abilities to respond to a contested environment will provide an incremental improvement in the resilience of the space mission. Clearly, survivability needs to be designed into the space segment, not left as an afterthought to be handled by the ground element.

Objectives

We were tasked by Air Force Space Command (AFSPC) to assess and modify some of the elements of DOTMLPF-P to improve space mission resilience. We had two principal objectives. The first is to identify and recommend near-term (i.e., "low-hanging fruit"), mid-term, and far-term DOTMLPF-P actions that will help improve space mission resilience, independent of the space segment architecture. The recommendations would consider ease of implementation and cost to implement, as well as overall improvements to the space mission resilience. The results of that research are the subject of this report. The second objective is to identify and develop an analytical methodology to assess how modifications to elements of DOTMLPF-P would improve the resilience of the overall space mission. The resulting methodology and analytical tool is the subject of a companion report.[15]

Scope

This report addresses improvements to five DOTMLPF-P elements that were deemed to present the most opportunities for actionable recommendations. The five elements were selected with the research sponsor and are defined as follows:

- doctrine: the way we fight, including tactics, techniques, and procedures (TTP)
- organization: how we organize to fight, including command and control (C2)
- training: how we prepare to fight tactically (basic training to advanced individual training, unit training, joint exercises)
- personnel: availability of qualified people for peacetime, wartime, and various contingency operations, including personnel performance
- facilities: real property, installations, and industrial facilities (e.g., ground control center).

The recommendations presented are independent of the space mission (e.g., environmental monitoring versus missile warning); however, they may need to be applied differently, based, for example, on mission criticality, with additional resilience needed for the most critical missions. Finally, we note that some of the recommendations provided in the report require more in-depth evaluations (including cost-benefit analysis) to determine how best to apply them—for example, use of government civilians versus contractors.

While the results of our research may apply to other space missions, we focused on force enhancement missions[16] because these capabilities are likely to be targeted by a potential adversary who wants to degrade the warfighting effectiveness and efficiency of our air, ground, and naval forces.

We also focused on the tactical or squadron level (i.e., space operations squadrons [SOPSs] and space warning squadrons [SWSs]), where C2 of Air Force satellites occurs, and on the

---

[15] Paul Dreyer, Krista S. Langeland, David Manheim, Gary McLeod, and George J. Nacouzi, *RAPAPORT (Resilience Assessment Process and Portfolio Option Reporting Tool): Background and Method*, Santa Monica, Calif.: RAND Corporation, RR-1169-AF, 2016.

[16] Force enhancement missions include positioning, navigation, and timing; military satellite communications; missile warning; environmental monitoring; and intelligence, surveillance, and reconnaissance.

operational level (i.e., the Joint Space Operations Center [JSpOC], which is responsible for operational employment of worldwide joint space forces).

We recognize that the space domain involves systems from other services, the intelligence community, civil agencies, allies, coalition partners, and commercial entities. Some of our recommendations for the Air Force will have applicability to these other space operators, but our focus was on AFSPC systems and, as noted above, its force enhancement missions. Similarly, we did not examine the cyber domain, which we understand will be more and more integrated with the space domain, with capabilities, vulnerabilities, and resilience of the two domains inherently linked.

Space Resilience

DoD has defined *resilience* in the context of space as

> [t]he ability of an architecture to support the functions necessary for mission success with higher probability, shorter periods of reduced capability, and across a wider range of scenarios, conditions, and threats, in spite of hostile action or adverse conditions.[17]

While this definition and others exist, for our purposes, enhancing resilience involves improving mission assurance in a contested environment. We take this broad view for the research reported here and the companion methodology report mentioned above.

However, because we are focused on the near term (less than three years), the options for enhancing resilience are more limited, as the space and ground segment are in place and are unlikely to change much in this time frame. While the space community currently lacks a taxonomy for discussing resilience,[18] there are a number of approaches being discussed to increase the resilience of an architecture. They include disaggregation, distribution, diversification, protection, proliferation, and deception.[19]

At the ground element, specifically the squadron level, enhancing "resilience" really devolves to space protection. The other five categories, as usually defined, apply to how one has already implemented the space and ground segments. Space protection, better described as self-protection for our purposes,[20] involves passive measures (e.g., anti-jam capability and nuclear hardening) and active measures (e.g., maneuverability) that the satellite can take to defend itself—often with the assistance of an operator—assuming that such capabilities already are

---

[17] DoD, "Space Policy," DoD Directive 3100.10, October 18, 2012.

[18] During preparation of the final report, the authors learned that DoD had released a taxonomy. See Office of the Assistance Secretary of Defense for Homeland Defense and Global Security, "Space Domain Mission Assurance: A Resilience Taxonomy," white paper, Washington D.C., September 2015.

[19] During preparation of the final report, the authors learned that the Air Force, following DoD's recent Strategic Portfolio Review for Space, is leaning away from disaggregation, a concept advocated in an AFSPC white paper released in August 2013, toward a broader theory informally known as *space protection*, and that disaggregation is now seen as just one element of a broader space-protection solution (Mike Gruss, "Disaggregation Giving Way to Broader Space Protection Strategy," *Space News*, website, April 26, 2015).

[20] We did not address offboard protection measures. That is usually considered part of defensive space control and is beyond the scope of this research.

resident on board the satellite. While JSpOC can take broader actions to enhance resilience than the squadron can, we primarily focused on those actions that enable the squadrons to better protect their space assets. Thus, while we use the term *resilience* throughout the report, when we discuss specific recommendations for the Air Force, they are primarily in the context of space protection.

Approach

The first step consisted of a comprehensive literature search in which we reviewed best practices used by other organizations and industry to implement resilience.[21] Because of the extensive nature of the review, we summarize the results in this report and present the details in an accompanying report.[22] We also examined U.S. civil organizations that operate satellites to identify practices that may be applicable to the Air Force.[23] However, much of the insight we developed about non-materiel aspects (i.e., DOTMLPF-P) of resilience for Air Force space operations was derived from meetings and discussions we held with subject matter experts (SMEs), operators at the SOPSs/SWSs and JSpOC, and other stakeholders.[24]

We then identified a number of shortcomings related to resilience, developed an initial set of mitigation options (based on all the discussions, literature reviews, and our own expertise), and met again with available stakeholders to get feedback on the candidate mitigation approaches. Given the interdependencies between the various DOTMLPF-P elements, an effective plan to improve space mission resilience required an overarching operational concept for space protection. A final list of actionable recommendations was then developed, based on this operational concept, to mitigate the identified resilience shortcomings. The recommendations are categorized in bins of near-term ("low-hanging fruit"), mid-term, and far-term implementations.[25]

Note that the findings and recommendation are discussed in terms of this operational concept, and not explicitly in terms of the individual elements of DOTMLPF-P. In some cases, it was not clear in which category a specific non-materiel finding or recommendation should appear (such as whether an item was an organization issue or a doctrine issue). The operational concept not only obviated the need to make that decision, but also provided a framework that provided operational value.

---

[21] Different organizations (e.g., the health industry) define *resilience* differently. It may be defined as redundancy, continuity of operations, or robustness. However, they have similar objectives.

[22] Krista S. Langeland, David Manheim, Gary McLeod, and George J. Nacouzi, *How Civil Institutions Build Resilience: Organizational Practices Derived from Academic Literature and Case Studies*, Santa Monica, Calif.: RAND Corporation, RR-1246-AF, 2016.

[23] They include the National Aeronautics and Space Administration (NASA), the National Oceanographic and Atmospheric Administration (NOAA), and the U.S. Geological Survey (USGS).

[24] These SMEs generally have ten or more years of space experience and hold positions of command authority or have responsibility for specific activities (e.g.. tactics development, training) and, thus, have current knowledge of the state of Air Force space operations.

[25] Due to time constraints, the developed methodology was not used to assess the recommendations made in this report.

Report Structure

The report is organized into six chapters and two appendixes. As noted earlier, there are two other accompanying reports: One presents the results of a comprehensive academic literature survey pertaining to how civil industries define and implement resilience,[26] and the second focuses on the resilience assessment methodology.[27]

Chapter 2 presents an overview of the results of the literature survey and case study analysis on best practices used by civil organizations to implement resilience, with the expectation that our findings would have applicability to the Air Force space operations community. Chapter 3 summarizes outcomes from the various discussions regarding space resilience we had with three civil space organizations: NASA, NOAA, and USGS. It discusses practices in civil space operations that may be applicable to the Air Force to improve resilience. Chapter 4 discusses resilience from an Air Force space operations perspective. It presents the shortfalls in current DOTMLPF-P elements that affect resilience and recommendations on how to mitigate them. This should be considered the key chapter of the report. Chapter 5 examines some of the mitigations listed in Chapter 4 that would be affected by a new space architecture that includes heavier reliance on international and commercial partners. Chapter 6 closes the report with some summary observations and recommendations.

Appendix A describes the assumptions, data sources used, and the basis for the rough order of magnitude (ROM) nonrecurring engineering (NRE) and recurring (REC) cost estimates for implementing the set of near-term, mid-term, and far-term mitigation options for improving space resilience that are presented in Chapter 4. Appendix B examines the findings and recommendations in Chapter 4 as applied to a specific force enhancement mission—namely, missile warning. Appendix B is provided under separate cover.[28]

---

[26] Langeland et al., 2016.

[27] Dreyer et al., 2016.

[28] Myron Hura, Gary McLeod, and George J. Nacouzi, *Enhancing Space Resilience Through Non-Materiel Means: Appendix B—Missile Warning Mission Case Study*, Santa Monica, Calif.: RAND Corporation, 2016, not available to the general public.

2. Resilience and Civil Institutions

Ensuring capable operations over time in the face of adversarial action, financial constraints, and severe weather prompts organizations, including municipalities, critical public services (e.g., hospitals), and large businesses, to develop plans for continuing to operate during and after a potential disaster. A review of the academic literature on resilience and related terms was conducted to determine how these entities define and assess resilience and prepare for, organize for, and respond to a threat event. The results from this literature review yielded insights and general guidance for assessing and building resilience. Specifically, the results from this review offer general insights that could be used by the Air Force space operations community to address resilience in a non-materiel manner. A general summary from this review is presented here, and a more detailed discussion can be found in a companion report.[29]

Resilience, as presented in the literature, is an attribute of a system that generally indicates its ability to maintain critical operations in the face of adverse disruptions. Beyond this general definition lie many variances based on community characteristics, threat environments, and overall operational goals. Such attributes as complexity, structure, training, and performance objectives determine how a community approaches resilience, while characteristics of the operational environment, including risk tolerance, scope of possible threats, and expected impact, indicate which metrics are appropriate for assessing resilience. Different communities have, therefore, developed unique concepts of and approaches to resilience, along with appropriate corresponding metrics. In the psychological community, resilience is demonstrated when an individual emerges from an adverse experience with increased psychological and emotional strength. The factors that make an individual psychologically resilient are often more subjective and attitude-related, and here the primary resilience metric is the emotional well-being of the individual.[30] In contrast, engineers characterize a structure as resilient based on its ability to avoid failure; factors that contribute to structural resilience include physical strength and robustness, and the ability to avoid structural failure is the primary resilience metric.[31] In ecological communities, the ability to adapt to new threats enables resilience of the entire community, and this flexibility is an important metric in evaluating these systems.

Based on these variances, enhancing resilience requires a varied approach that takes into account these community attributes and the operational environment. This chapter presents the approaches taken by three different types of communities to develop and maintain resilient operations. The discussion presented here illustrates methods for building resilience through withstanding an adverse event (impact avoidance and robustness), resilience through adaptation and flexibility, and resilience through recovery and restoration. Communities seeking to develop

---

[29] Langeland et al., 2016.

[30] Psychology Today, "Psych Basics: Resilience," website, New York, N.Y., undated.

[31] B. J. Jennings, E. D. Vugrin, and D. K. Belasich, "Resilience Certification for Commercial Buildings: A Study of Stakeholder Perspective," *Environment Systems and Decisions*, No. 1779, 2013.

more resilient operations can gain insight from the academic studies and experience provided in the literature. By identifying the methods and lessons derived from previous studies of similar operational environments and how they addressed resilience, similar operations can benefit from this insight. These approaches are broadly described in this chapter as *withstand*, *adapt*, and *recover*. Recognizing that any given organization will incorporate all three of these approaches into an overall resilience plan, this discussion seeks to highlight organizations that will be most likely to emphasize one of these approaches over another.

General Approaches for Building Resilient Operations

Impact Avoidance

Communities that have a low tolerance for risk or failure seek to withstand a potential degradation or disruption by avoiding impact entirely. These "hazardous" industries include, for example, air traffic control and nuclear power plant management. Hazardous industries are characterized by their unforgiving environment and severe consequences for mission failure,[32] and the resilience of a hazardous industry is determined by its ability to avoid degradation. This depends on its reliability or robustness when the mission being considered cannot tolerate any degradation, error, or failure.[33] These organizations often prioritize performance above, for example, profitability, timeliness, and efficiency.

Hazardous organizations that demonstrate a sustained ability to avoid failure are referred to in the literature as high-reliability organizations (HROs).[34] These organizations are often challenged by high levels of complexity, advanced technology, and tightly integrated systems. HROs develop methods for addressing these challenges and illustrate general lessons for high-risk industries seeking to mitigate impact. Below we summarize lessons from the academic literature on HROs and discuss lessons learned from a case study of hazardous industry failure—specifically, the Fukushima Daiichi nuclear disaster in 2011.

Error reporting and risk assessment emerged as two key components of a resilient industry in building HROs where risk tolerance is low.[35] To address the first component, error reporting, industries with catastrophic consequences for failure should have a different calculus for addressing resilience than those that can tolerate disruption. These industries cannot rely on trial

---

[32] K. Weick, K. Sutcliffe, and D. Obstfeld, "Organizing for High Reliability: Processes of Collective Mindfulness," in R. S. Sutton and B. M. Staw, eds., *Research in Organizational Behavior*, Vol. 1, Greenwich, Conn.: Jai Press, 1999, pp. 81–123; Todd R. LaPorte, "High Reliability Organizations: Unlikely, Demanding, and at Risk," *Journal of Contingencies and Crisis Management,* Vol. 4, No. 2, June 1996.

[33] K. Roberts, "Some Characteristics of One Type of High Reliability Organization," *Organization Science,* Vol. I, Issue 2, 1990.

[34] Roberts, 1990; Laporte, 1996.

[35] Linda T. Kohn, Janet M. Corrigan, and Molla S. Donaldson, eds., *To Err Is Human: Building a Safer Health System*, Washington, D.C.: National Academy Press, 1999; J. S. Weissman, C. L. Annas, A. M. Epstein, et al., "Error Reporting and Disclosure Systems: Views from Hospital Leaders," *JAMA,* Vol. 293, 2005; Z. R. Wolf and R. G. Hughes, "Error Reporting and Disclosure," Chapter 35 in R. G. Hughes, ed., *Patient Safety and Quality: An Evidence-Based Handbook for Nurses,* Rockville, Md.: Agency for Healthcare Research and Quality, 2008.

and error to boost resilience due to the high cost of error, but they can increase resilience by focusing on and learning from failure and near misses. To accomplish this, the organizational structure must support a culture that encourages reporting and communication. This culture is characterized in Weick, Sutcliffe, and Obstfeld as "collective mindfulness" that emphasizes not only observation but also subsequent action; this is presented as a key contributor to the success of HROs.[36]

To address the second key component, risk assessment, a comprehensive understanding of risks should be prioritized. Developing a complete understanding of the full range of risks is an intractable goal, and hardening against the highly improbable may be cost-prohibitive. However, emphasizing an understanding of the *highest-impact* risks is particularly important for hazardous organizations. An accurate assessment of the full range of risks permits the organization to prepare for the worst, not the most probable. This concept is illustrated vividly in the case of the Fukushima Daiichi nuclear disaster. In this case, an inaccurate assessment of possible tsunami magnitude and risk resulted in an avalanche of failures that eventually resulted in catastrophic radiation release.[37] This case study illustrates the dire consequences that can occur when the impacts of possible threats are not fully understood or accounted for.

In addition to accurate risk assessments, a culture of communication and collaboration significantly contributes to the robustness of the HRO. Collaboration requires reliable communication infrastructure that includes a clear reporting system, allows appropriate information flow, and facilitates shared mission awareness and shared mission goals. Encouraging the reporting of potential errors or near misses and incorporating this information into future training and response plans is a key activity for building resilience in a high-risk environment. Coupled with this reporting system is a hierarchical structure with sharp lines of accountability, giving each party a shared responsibility for the mission objective.

The skills and ability of personnel to make decisions are especially important in the high-risk environment discussed here. While training is always important in any environment, because of the complexity of the systems used in a typical HRO, detailed training on specific systems should particularly be emphasized. Personnel with detailed technical knowledge of the systems being used will be more able to identify potential mitigations when threatened with potential performance degradation or failure. Due to the high level of integration between these systems, however, personnel must also have general background and knowledge of these systems to ensure flexibility during emergency operations. Successful HROs find a balance between these seemingly paradoxical requirements.

Overall, reporting and learning from errors and near misses is a prominent trait of HROs. A collaborative culture that adjusts dynamically to changing circumstances and new information facilitates learning from these events while enabling a shared mission awareness among all personnel.

---

[36] Weick, Sutcliffe, and Obstfeld, 1999.

[37] M. Hirano, T. Yonomoto, M. Ishigaki, N. Watanabe, Y. Maruyama, Y. Sibamoto, et al., "Insights from Review and Analysis of the Fukushima Dai-Ichi Accident," *Journal of Nuclear Science and Technology*, Vol. 49, No. 1, January 2012; TEPCO, "Fukushima Nuclear Accident Analysis Report," June 20, 2012; J. M. Acton and M. Hibbs, "Why Fukushima Was Preventable," Carnegie Endowment for International Peace, March 6, 2012.

Adaptation and Flexibility

Adaptation is the most appropriate approach for those industries that have more flexible tolerances for small disruptions, and many customer-facing businesses (such as power and other utilities) fall into this category, as they continually respond to the dynamic needs of their customer base. These industries prioritize the ability to operate through changes and disruptions, maintaining critical capabilities during and after a disruption. The primary metric for these industries is flexibility, or the ability to evolve to accommodate changing circumstances. Business operations that require continuous operations in a dynamic environment exemplify this approach toward resilience, one that seeks a balance between efficiency and reliability, and our review references literature on business management and operations to determine how these organizations support resilience through flexibility.

Business operations seek to ensure far-term mission success and maximum profitability by finding an optimal balance between efficiency of operations and minimization of vulnerability to disruption, essentially balancing resilience with day-to-day productivity. Supply chain management exemplifies such balance.[38] Resilience in a supply chain is achieved by optimizing productivity while minimizing risk. To assess vulnerability and corresponding resilience, businesses will often list risks and impacts and then run comparative analyses of each.[39] While this list cannot feasibly be exhaustive in most cases, strategies for resilience against known threats are pursued with the assumption that these mitigation strategies, while specific to one type of threat, also harden the system against other unforeseen threats.

One important method for achieving resilience is implementing measures to increase information availability when practical and appropriate. Not only does increased visibility expedite the detection of potential disruptions and enable impact mitigation from these disruptions as they occur, but this increased situational awareness facilitates the identification of inefficiencies in the overall process. The increase in information made available to personnel needs to be balanced with the ability of the personnel to ingest this information. A flood of information is not likely to be useful and could even damage resilience efforts, so a balance must be achieved. In general, this increased information availability helps to mitigate impact even in the absence of a specific response plan, since all involved parties have the requisite knowledge to make informed operational decisions.[40]

Designing processes and operations for flexibility is another key method for building resilience. This method entails the development of a proactive risk management strategy that requires a dynamic assessment of possible exposure to circumstances that could impact

---

[38] Maria Jesus Saenz and Elena Revilla, "Creating More Resilient Supply Chains," *MIT Sloan Management Review,* Summer 2014; Supply Chain Risk Leadership Council, "SCRLC Emerging Risks in the Supply Chain 2013," white paper, 2013.

[39] SCRLC Maturity Model Team, "SCRLC Supply Chain Risk Management Maturity Model," interactive spreadsheet, April 2, 2013.

[40] Saenz and Revilla, 2014; Deloitte, "Supply Chain Resilience: A Risk Intelligent Approach to Managing Global Supply Chains," 2012; Kelly Marchese and Jerry O'Dwyer, "From Risk to Resilience: Using Analytics and Visualization to Reduce Supply Chain Vulnerability," *Deloitte Review,* Issue 14, 2014.

capability.[41] By redesigning processes with responsiveness and flexibility in mind, it is possible to build a dynamic culture that is able to respond more effectively.[42] The design of these processes should consider approaches that include increasing redundancy, standardizing processes, and disaggregating integrated operations. Further, the organizational structure and culture can contribute to the flexibility of operations by fostering continuous communication and distributing decisionmaking power, as well as creating passion for work and training for disruption.[43]

Successful business practices, and supply chain management practices in particular, achieve a balance between reducing vulnerability and maintaining efficient operations by utilizing these approaches. The flexibility afforded through information flow, shared mission awareness, and processes that were redesigned with flexibility and adaptability in mind is key to mission resilience for a dynamic environment characterized by moderate risk. For organizations that require the ability to operate through a disruption or degradation and to adapt accordingly, emphasis should be placed on enhancing situational awareness and flexibility. This can often be achieved through increased information flow and distribution of decisionmaking power, though this must be balanced appropriately with consideration of the efficiency and effectiveness of daily operations.

Recovery and Restoration

In contrast with HROs that strive to be failure-resistant and business supply chains that are intended to operate through threat scenarios, some communities with high-priority missions instead emphasize the enabling of rapid recovery immediately following impact rather than avoiding or accommodating this impact. The reasons for this may include the degree of difficulty and expense required for hardening the organization to all possible threats, the large number of possible impacts that would need to be anticipated in order to avoid impact, and the sheer number of facilities that would need to be hardened to secure the entire system. In the wake of a disrupting event with high impact, such as a natural disaster, some high-priority communities strive for rapid recovery of mission-critical capabilities even as they follow procedures to minimize impact. These types of organizations emphasize the ability to recover as a basic resilience metric; example organizations include hospitals and electric power utilities. The recovery of basic capabilities in a hospital during and following a natural disaster will save lives. The restoration of electricity following a power outage will not only facilitate ongoing recovery efforts, but it could also prevent a larger system collapse and the severe economic impacts of such a blackout.[44] While all possible efforts are made to prevent and mitigate impact, optimizing

---

[41] Saenz and Revilla, 2014; Marchese and O'Dwyer, 2014.

[42] Yossi Sheffi, "Building A Resilient Supply Chain," *Harvard Business Review Supply Chain Strategy Newsletter,* October 2005.

[43] Deloitte, 2012; Sheffi, 2014; Saenz and Revilla, 2014.

[44] A. Bernstein, D. Bienstock, D. Hay, M. Uzunoglu, and G. Zussman, "Power Grid Vulnerability to Geographically Correlated Failures—Analysis and Control Implications," *INFOCOM Proceedings,* IEEE, 2014; Electric Consumer Research Council, "The Economic Impacts of the August 2003 Blackout," February 2004.

recovery efforts is central to improving resilience for critical missions in this type of organization.

Recovery operations can involve both evacuation to alternate facilities and the restoration of current infrastructure and/or facilities. For the first approach, evacuation procedures in response to a threat event, full-scale exercises conducted prior to threat impact can reveal more efficient methods and time-saving measures that can be incorporated into a periodically edited emergency response plan.[45] This response plan is a key component of resilience operations, and incorporating lessons learned from response planning exercises can significantly increase the efficiency and efficacy of future response efforts.

The second approach is repairing and restoring infrastructure. During disaster response, time is critical, and the speed with which the source of the failure can be identified and addressed is the primary metric for resilience. Practices used by electric utilities to minimize downtime and expedite recovery from an outage yield key guidelines for achieving resilience through rapid recovery. These guidelines include having experienced personnel on call, supporting shared mission awareness among all personnel, and establishing coordinated reporting procedures.[46] Experienced personnel can more rapidly identify sources of failure and respond, resulting in significant reductions in loss. Coordination similarly minimizes response time by avoiding duplication of efforts and mitigating confusion if the activities need to divert from the response plan.

In both evacuation and restoration operations, dynamic emergency response plans that incorporate new information and are well-rehearsed are vital components of a successful recovery effort.[47] Such a response plan will minimize response time by allowing efficient allocation of resources and strategic prioritization of efforts. In addition, the availability of well-qualified personnel who can be quickly utilized is key for operational success. Organizations that strive to recover from a threat impact quickly need to both prepare and continuously update disaster response plans based on results from testing and exercises and keep sufficient numbers of qualified personnel in a position to act in case of disaster.

Potential Applications to the Space Operations Community

Each of the concepts presented above provides some specific insight into methods for ensuring resilience for different mission goals: avoid risk, operate through the threat event, and recover. Hazardous industries in particular may significantly increase resilience by learning from errors and developing detailed yet dynamic response plans. Supply chain management seeks to develop resilience through enhanced situational awareness and distributed and flexible decisionmaking. Recovery operations often can improve resilience by incorporating lessons from testing and exercises and assuring availability of skilled personnel on call. These specific

---

[45] C. Verni, "A Hospital System's Response to a Hurricane Offers Lessons, Including the Need for Mandatory Interfacility Drills," *Health Affairs,* Vol. 31, No. 8, 2012.

[46] J. P. Smith and Gulfport CARRI team, "Organizational Resilience: Mississippi as a Case Study," Gulfport Resilience Essay of the Community and Regional Resilience Institute, March 2013.

[47] Smith and Gulfport CARRI team, 2013; Verni, 2012.

concepts can provide some guidance tailored for a particular industry type or mission objective. Yet, while each of these approaches to resilience may utilize unique practices and methods, common themes were repeated throughout the literature. These common themes in particular may offer guidance for increasing resilience in the diverse and complex space community.

Information-sharing and shared awareness of the mission will increase the efficiency and effectiveness of operations both during and following a threat event. Implementing organizational structures and building internal cultures that support and encourage information flow and situation awareness are shown in the literature and case study reports to optimize operations and personnel performance during and following a threat event or disruption. The Air Force could enhance resilience in the space operations community by finding ways to more effectively share information among personnel in different roles and at different levels of authority. This information-sharing must be balanced, however, with effectiveness of operations and appropriate designation of decisionmaking authority. Overwhelming personnel with unneeded information could inhibit effective operations, thereby decreasing resilience; allowing personnel to make decisions based on partial information will also negatively impact operations. Organizations must, therefore, carefully evaluate how they can make as much information as possible available to decisionmakers without overwhelming them with extraneous data.

Clear reporting structures and cultures that support error reporting will allow an organization to develop more resilient operations by incorporating lessons from previous errors. Information flow is supported by a clear reporting structure that not only supports integrated communication but also builds accountability. Establishing well-defined reporting procedures is shown to maximize the efficiency and timeliness of operations in each mission type discussed here, and developing a culture that supports the reporting of failures and near misses is a key element of this reporting structure. Similarly, the Air Force space operations community could build resilience by implementing measures to encourage error reporting free of the threat of reprimand and establishing ways to incorporate lessons from these errors in real time.

An appropriate balance between flexible personnel with distributed decisionmaking and specialized personnel with centralized decisionmaking will support the ability to observe and react appropriately. The ability to act outside of established response plans is key to building general resilience against unanticipated threats and impacts. Qualified personnel will have the ability to adapt responses in real time, and this ability needs to be accompanied by appropriate decisionmaking authority. Distributing this authority during a threat response will enable swift response and action. Training programs in the Air Force could be adapted to ensure that operators have the expertise required to react in real time. However, distributed decisionmaking is more challenging for a large organization, and the Air Force may find that incorporating a more distributed authority may compromise the efficiency of its day-to-day operations. For this reason, a balance needs to be achieved between the flexibility of the personnel and their level of specialization; an inflexible hierarchy is brittle, but full authority being granted to all personnel is also brittle. The location of this balance is often challenging to identify and is determined by a variety of factors, including organizational size, operational requirements, and personnel skill level. Maximizing expertise and flexibility within this balance will support the ability to observe and react appropriately, in the manner of organizations that demonstrate collective mindfulness.

Accurate risk assessment methods will facilitate better design and planning. While the full spectrum of possible impacts and risks may be impossible to capture, an accurate assessment of risk and failure tolerance will facilitate resource allocation and investment decisions. The Air Force could enhance its space mission resilience by investing resources in risk assessment modeling and fault tolerance testing of critical ground systems that support space operations. In addition, consideration of mission and operation type can inform fault tolerance requirements. For example, systems used for risk-averse missions should be designed to withstand maximum failure and avoid impact, while systems used to operate through a threat with some degradation in capability should be designed for maximum flexibility.

Training for specific threats while maintaining flexibility in response procedures is a challenge, but meeting this challenge will allow an organization to address both specific and general threats. Developing appropriate training programs is crucial to ensuring an effective response to a threat event, and detailed exercises that address specific and known threats are crucial. However, unanticipated threats require personnel flexibility to respond outside of programmed procedures, and this presents a paradoxical challenge to operational managers, requiring compromise between efficient day-to-day operations and maximizing flexibility for disaster response. The Air Force could benefit from more frequent and more detailed exercises, but ensuring that ad hoc response capabilities are developed needs to be similarly prioritized.

Summary

The techniques presented in this literature review are intended to facilitate the identification of appropriate steps that a variety of organizations and communities, including the Air Force, could take to increase its resilience, particularly with limited resources. The lessons learned are not intended to be comprehensive, but instead provide some general guidelines for optimizing resource investment while assuring continued and successful operations. An organization can identify which of the measures summarized here are most appropriate and accessible based on resource availability, expected threats, organizational and operational restrictions, and mission requirements. Implementing these measures may allow these organizations to take significant steps toward sustaining mission-critical capabilities.

3. Resilience and U.S. Government Civil Space Agencies

As discussed in Chapter 2, different communities have different definitions of resilience, along with appropriate corresponding metrics based on community attributes, threat environments, and overall operational goals. This chapter examines U.S. civil space agencies' practices that affect resilience, in the context of civil community attributes and operational environment, to identify those that may be applicable to the Air Force.

We begin the chapter with a discussion of policy, which drives the civil space agencies' attributes and operational environment. We then identify specific operational practices that describe how the civil space agencies are postured for resilience. In both policy and operations, we find that international data sharing drives increased focus on information and communications technology (ICT) architecture and that the basis for resilience is driven primarily by continuity of operations (COOP).

As a reminder, three civil space agencies operate satellites: NASA, NOAA (which is part of the U.S. Department of Commerce [DoC]), and USGS (which is part of the U.S. Department of the Interior [DOI]).[48]

Civil Policy Considerations

The civil space programs differ from national security space (NSS) programs in terms of their policy objectives. Among the civil space programs, we found that NOAA's space-based weather collection mission was most similar to the Air Force's space force enhancement mission because of its emphasis on time-critical, assured delivery.

Full and Open Access

Civil space operators provide "full, open, and timely access to government environmental data" to international users, as mandated by the *National Space Policy of the United States of America*.[49] USGS provides access to all Landsat data at no cost.[50] These policies expose the civil space ICT architecture in a fundamentally different manner than DoD. Many of NOAA's international relationships are with countries not necessarily friendly with the United States (e.g., NOAA has 12 active projects that involve Cuba). Major partners in the very popular international Search and Rescue Satellite Aided Tracking (SARSAT) system, which dates back to 1981, include the United States, Canada, France, and Russia.

---

[48] NOAA operates geostationary weather satellites and low Earth orbit polar weather satellites, including the Defense Meteorological Satellite Program on behalf of the Air Force. USGS operates the Landsat missions. NASA acquires and launches all U.S. civil scientific satellites, and it owns and operates the largest and most diverse fleet of the three agencies.

[49] White House, *National Space Policy of the United States of America*, Washington, D.C., June 28, 2010.

[50] United States Geological Survey, "Landsat—A Global Land-Imaging Mission," USGS Fact Sheet 2012-3072, May 2013.

Rapid Delivery

NOAA must deliver satellite data quickly and reliably because the value of the data degrades very rapidly. Simply put, "Weather data has a rotten shelf life."[51] The NOAA follow-on civil weather satellite program, Joint Polar Satellite System, has a 96-minute data delivery latency Key Performance Parameter, which includes collection, downlink, processing, and worldwide open dissemination.

Continuity of Operations

NOAA's assured delivery of satellite weather data is viewed as critical at the department level; provision of satellite weather data products is one of four Primary Mission Essential Functions (PMEFs) assigned to DoC in the *National Continuity Policy* and *Assignment of Emergency Preparedness Responsibilities*.[52] Any issue that causes a delay of over two hours in the delivery of weather data results in notification of the Secretary of Commerce. A senior government official pointed out that weather forecast consumers (essentially, every United States citizen) form a visible and expansive constituency from which feedback for poor performance is received quickly and loudly.[53]

Security Classification

U.S. government security clearances are not the norm for civil space operators. NASA's Goddard Space Flight Center and the NOAA Satellite Operations Facility (NSOF) had at least one person within the operations center at all times with a U.S. government secret security clearance.[54] Access to secure communications and computing is available at the NSOF, although it is not integrated into the operations floor.

Civil Practices

We identified specific behaviors across the civil community that could be advantageous when operating satellites in a contested space environment.

Information

Civil agencies do not maintain their own threat condition, and none of those surveyed currently had a formal mechanism for disseminating intentional threat information. As an example, USGS relies on a small cadre with security clearances to maintain awareness of threat

---

[51] NOAA, discussions regarding NOAA space operations and resilience with Office of Satellite and Product Operations staff, Suitland, Md., September 4, 2014a.

[52] White House, *National Continuity Policy*, National Security and Homeland Security Presidential Directive, NSPD-51/HSPD-20, Washington, D.C., May 4, 2007; White House, *Assignment of Emergency Preparedness Responsibilities*, Executive Order 12656, Washington, D.C., November 18, 1988.

[53] NOAA, 2014a.

[54] NOAA, 2014a, and NASA, discussion regarding NASA space operations and resilience with Goddard Space Flight Center staff, Greenbelt, Md., August 29, 2014a.

information. Members of the cleared cadre reported difficulty balancing their fellow operators' needs for specificity with classification restrictions.[55] The issue of how to report actionable warning without disclosing sources and methods is in active discussion at NASA.[56] Civil space agencies are more exposed to cyber threats than national security space systems because of their full and open access policies. For example, their dissemination systems exist on the open Internet.

Both NASA and NOAA are investigating the possibility of introducing predictive analytics to satellite operations. Loosely based around the Air Force Research Laboratory's "Satellite as a Sensor" program,[57] these predictive analytics concepts run all spacecraft telemetry through a machine learning system (typically, an artificial neural network) to characterize nominal operations and identify off-nominal conditions.[58] These tools enable not only more timely detection of anomalies, based on the state of health data from each satellite, but also detection of anomalies that an operator may miss. In the industrial automation sector, this category of techniques is categorized as "predictive maintenance" and has demonstrated real return on investment across industries including automotive, energy, and electronics production.[59] Predictive maintenance applications use analytics techniques to identify patterns of failures based on minimally filtered data from across the enterprise.

The NASA and NOAA efforts, while nascent, look to capitalize on both satellite-specific behavior models and enterprise-wide behavior models. Aggregating data from discrete satellite systems into enterprise-wide models may permit insight into systemic effects, such as space weather or cyber disruptions.

Organization and Tactics

None of the civil agencies designate a position on their operations floors that addresses space protection issues. Currently, it is the responsibility of mission directors to understand the threat environment, assess whether anomalies may be a result of intentional acts, and respond accordingly. Because of their criticality, it may be worthwhile to assign such functions initially to a space protection lead (i.e., make them the secondary duty of one of the operators) and then later to a separate space protection position, as the threats mature. However, this would not absolve the other operators from also being on the outlook for anomalies and bringing them to the attention of the space protection lead and mission director.

---

[55] USGS, discussions regarding USGS space operations and resilience with Flight Systems staff, Greenbelt, Md., September 29, 2014.

[56] NASA, 2014a.

[57] Air Force Research Laboratory, "Innovation: Threat Detection, Validation, and Mitigation Tool for Counterspace and Space Situational Awareness (SSA) Operations," SBIR Topic No. AF06-283, Wright-Patterson AFB, Oh., undated.

[58] C. R. Tschan and C. L. Bowman, *Development of the Defensive Counterspace Test Bed (DTB), Volume 1—Sensors and Detection*, TOR-2004(1187)-2, El Segundo, Calif.: Aerospace Corporation, 2004.

[59] Stephanie Diamond and Anuj Marfatia, *Predictive Maintenance for Dummies*, Hoboken, N.J.: John Wiley and Sons, 2013, pp. 12–15.

As it is, NOAA routinely practices debris avoidance in its low Earth orbit (see more in the "Training" section below), although it has not explicitly developed tactics to avoid a deliberate adversary. NASA Johnson Space Center has a ground controller console position for human spaceflight missions whose responsibility is to ensure proper functioning of the cyber networks.[60]

Command and Control

NOAA operates all of its satellites at the NSOF and maintains capability for fully operational failover to a designated backup facility. NOAA has the advantage of having all its missions under one roof, which increases shared situational awareness between its constellations. However, the processes for space situation information sharing are organic, not as a result of designated command authority. NOAA reported that it has an easier time reporting classified information to higher authority than receiving classified instructions from higher authority.

ICT architecture is a point of pride for NOAA. Due to the full and open access policy for space-based weather data, close attention has been paid to separation of C2 from data dissemination networks. As characterized by a NOAA representative, the NOAA information architecture is engineered to "control the keys to the car."[61]

NASA does not have a centralized control facility but instead distributes control across its centers. The most prominent centers for satellite operations are the Jet Propulsion Laboratory (for interplanetary missions), Goddard Space Flight Center (for relay and earth science missions), and Johnson Space Center (for manned flights). Each center has its own operations tempo and culture, as dictated by the mission.

Training

In general, civil space agencies field seasoned and experienced flight crews with minimal turnover. For example, NOAA reported that its operators generally have ten or more years of experience on console. Rather than inexperience, NOAA's concern is with the number of retirement-eligible employees in its operator ranks. Each of NOAA's constellations operates and maintains its own formal training and certification process for operators.[62] Many USGS operators have been flying Landsat 7 since its launch in 1999.[63]

As required by its science mission, NOAA's Suomi National Polar-orbiting Partnership (SNPP) satellite was launched into a sun-synchronous orbit, 833 kilometers mean altitude, and 10:00 a.m. local time of ascending node. The debris field from the Chinese Fengyun 1C weather satellite, destroyed by a Chinese anti-satellite weapons test on January 11, 2007, is currently

---

[60] NASA, discussions regarding NASA space operations and resilience with Johnson Space Center's Vehicle Integration Office staff, Houston, Tex., September 23, 2014b.

[61] NOAA, 2014a. Here, "keys" refers to C2, and "car" refers to the satellite.

[62] NOAA, discussions regarding NOAA space operations and resilience with Office of Satellite and Product Operations staff, Suitland, Md., September 10, 2014b.

[63] USGS, 2014.

transiting through the SNPP orbit. SNPP is NOAA's first maneuverable weather satellite and has maneuvered 18 times in its first 35 months for debris avoidance.[64]

Frequent maneuvering led to an unintended but favorable consequence: The SNPP crew and NSOF are trained and experienced in predictive avoidance. NOAA reports that its planning cycle and process for maneuvering is streamlined and efficient. While it is understood that the case of a directed adversary has a different precipitating event and timeline, valuable skills in maneuver planning, execution, and return to operations have been learned by NOAA through constant exercise. As the organization adapted to its environment, it learned skills that make it more resilient in a threat environment.

Ground support for the human spaceflight program is arranged as a hierarchy, with the flight director as the primary decisionmaker. The flight director sits in the front control room, where summary consoles for each of the primary mission systems are arrayed. Behind each console, there are teams in back rooms that monitor and control the subsystems. NASA requires time on console at particular stations and cross training at multiple stations to move up in the hierarchy (from the back rooms to the front room). NASA runs a formal training and certification process for its operators.[65]

NOAA routinely exercises COOP, including full failover to backup, in accordance with its PMEF responsibilities under the National Continuity Plan. The Johnson Space Center has implemented failover capability for the International Space Station mission control to Marshall Space Flight Center.

Personnel

No clear consensus on government/contractor force mix for operations positions emerged from the civil space community. Like the Air Force, NOAA employs only government personnel on console at NSOF. But USGS operators are all contractors, and NASA uses a mix. For the human spaceflight program, decision authority is vested in the flight director, a government employee, but all lower positions in the hierarchy can be government or contractor personnel.

Summary

We found that the following civil practices may increase resilience:

- Civil agencies require years of console experience for their operators. While the agencies use a variety of government/contractor mix strategies to assure continuity, the result is console operator cadres with an average of ten or more years of experience.
- NOAA has improved its resilience, although indirectly, by following strict COOP requirements, including a PMEF to provide space-based weather data. In addition, practice makes perfect, and constant practice leads to organizational learning and optimization. Specifically, NOAA's frequent maneuvering of SNPP serves as valuable training for space resilience.

---

[64] NOAA, 2014b.

[65] NASA, 2014b.

- Civil agencies are introducing predictive analytics to recognize satellite anomalies. While predictive analytics were originally demonstrated for satellite applications in the Air Force, they are now widely propagated across the business landscape, together with readily available commercial and open-source tools.

4. Resilience and Air Force Space Operations

Chapter 2 examined resilience from the perspective of civil institutions and highlighted practices that enhance resilience. Similarly, Chapter 3 examined resilience from the perspective of U.S. government civil space organizations. In this chapter, our focus is on Air Force space operations, principally the space operations squadrons (SOPSs) and space warning squadrons (SWSs) that operate satellites performing force enhancement missions[66] (i.e., the tactical level) and JSpOC, which is at the operational level.

We identified the shortfalls in current DOTMLPF-P elements that affect resilience and made recommendations on how to mitigate them using the following approach. We met with a number of Air Force space organizations[67] to better understand how the current DOTMLPF-P elements are affecting the overall space mission resilience. We heard what the current conditions are from SMEs[68] and what the Air Force is doing, if anything, to improve resilience. We collected these observations,[69] determined how they affected mission resilience, and assessed and refined many of the suggestions they provided. In addition, we added a number of our own options for improvement, and we reviewed resilience-enhancing practices identified in Chapters 2 and 3 for their applicability to the military space domain. At the end of the chapter, we present two detailed sets of recommendations: a list of low-cost near-term recommendations and a list of more robust, more expensive, far-term recommendations.[70]

Readers who are not familiar with the organization, C2 structures, and other ground elements that support space operations may refer to Joint Publication 3-14, *Space Operations*, and Air Force Doctrine Annex 3-14, *Space Operations*.[71]

---

[66] Force enhancement missions include positioning, navigation, and timing; military satellite communications; missile warning; environmental monitoring; and intelligence, surveillance, and reconnaissance.

[67] 50th Operations Group, Schriever AFB, Colo. (50 OG, 2014); 460th Operations Group, Buckley AFB, Colo. (460 OG, 2014); 11 Space Warning Squadron, Schriever AFB, Colo. (11 SWS, 2014); 614th Air and Space Operations Center, Vandenberg, AFB, Calif. (614 AOC, 2014); Air Force Space Command, Peterson AFB (AFSPC, 2014).

[68] The SMEs generally have ten or more years of space experience and hold positions of command authority or have responsibility for specific activities (e.g., tactics development, training) and, thus, have current knowledge of the state of Air Force space operations.

[69] These observations were consistent across the various space organizations, mainly because AFSPC has little activity under way (other than studies) regarding resilience and the force enhancement missions. At the beginning of the project, we heard senior AFSPC leaders say that it is time to stop admiring the problem and to start taking action.

[70] The results presented here are general and can be applied to all the force enhancement missions. Appendix B provides recommendations specific to the missile-warning mission. Appendix B is provided under separate cover.

[71] Joint Staff, *Space Operations*, Joint Publication 3-14, Washington, D.C., May 29, 2013; United States Air Force, *Space Operations*, Doctrine Annex 3-14, Curtis E. LeMay Center for Doctrine Development and Education, Maxwell AFB, Ala., June 19, 2012.

Operational Concept

Because of the many interdependencies among the DOTMLPF-P elements, rather than presenting our findings for each element, we found it more meaningful to organize them around a unifying operational concept—namely, "enhancing the capability of space operators to respond, in a timely and effective manner, to adversary counterspace actions." Implicit in this concept is that space operators will act to protect their systems; thus, at this level, "resilience" essentially devolves to "space protection," as discussed in Chapter 1.

To accomplish the space protection operational concept, operators need

- actionable information
- appropriate organization and tactics
- dynamic command and control.

These should all be supported by

- appropriate tools and decision aids
- relevant training and exercises
- qualified personnel.

These components will provide an end-to-end functional capability. If any are missing or degraded, execution of the operational concept will also be degraded or ineffective. The rest of the chapter is organized around the components of this operational concept, except for tools and decision aids, which are discussed as needed in each of the other five components. Note that we developed this operational concept because we were not aware of an existing CONOPS that addresses space protection from a tactical and operational perspective.

In the next five sections, we present our findings: We present our observations for each component, discuss their impact on resilience (at which point they become shortfalls), and suggest mitigations that address these shortfalls. In many cases, the potential mitigations are general recommendations and can be further delineated into actions that we call implementation options. We place the option in the near-term category if it is likely that it can be implemented within one year. This generally means that an option is relatively easy to implement (i.e., the amount of activity is modest and not beyond the usual range of duties) and the organization does not need higher-level approval. Generally, the cost is also low, but that will be discussed in a later section. Options are placed in the mid-term category if it is likely that they can be implemented within one to three years. Generally, these options require higher-level approval and may require more in-depth evaluation or assessment (e.g., defining requirements for system upgrades). Options in the far term will likely take three years or longer to implement because they require high-level approval, involve system development, or recommend additional manning (which is a difficult proposition with today's budget constraints). Note that some of the far-term options are also reliant on mid-term activities—e.g., defining requirements.

Findings: Information

Space operators need information to both recognize and properly respond to potential hostile activities. They also need information to rule out other external sources of anomalies, such as unintentional electromagnetic interference, space weather effects, and orbital debris. We have four observations that we believe have an impact on resilience.

Space Order of Battle

First, we found that space OB development is the responsibility of the National Air and Space Intelligence Center (NASIC), a science and technology intelligence center. Normally, OB development is performed at joint intelligence operations centers (JIOCs), which support each combatant command (COCOM). The current situation arose when U.S. Space Command was disestablished, and the space mission was transferred to U.S. Strategic Command (USSTRATCOM). At the same time, space OB development, which was performed by a group of government civilians in the combined intelligence center (CIC),[72] was transferred to U.S. Strategic Command's Joint Intelligence Center (STRATJIC). USSTRATCOM later transferred the responsibility to NASIC because of the expertise resident at that organization. We argue that, with USSTRATCOM's creation of functional components for day-to-day operational planning and execution of its diverse missions, responsibility for OB development, an operational intelligence function, should also be the responsibility of these operational functional components. Specifically, we recommend that space OB responsibility should be transferred from NASIC to the Joint Functional Component Command for Space (JFCC Space), and, in particular, to the JSpOC Intelligence, Surveillance, and Reconnaissance Division (ISRD). This would not only allow JSpOC ISRD to provide JFCC Space with more complete and timely operational intelligence support, but it would also complement one of JSpOC's principal responsibilities—space situational awareness. Further, it will enable JSpOC ISRD to provide more complete and timely operational intelligence to the SOPSs/SWSs.

We see no near-term implementation options. In the mid-term, JSpOC ISRD could take "responsibility" for OB development and dedicate staff to interface with NASIC staff, with NASIC still performing the bulk of the work. In the far term, JSPOC ISRD could establish an initial cadre of government civilians (three to six)[73] with extensive knowledge of U.S. and foreign space capabilities at JSPOC ISRD to maintain the space OB. This would partially replicate the OB development capability resident at the former CIC.

Limited Intelligence at SOPS/SWS

Second, we found that there is limited to no access at the SOPSs/SWSs to potential adversary counterspace force posture. Without this information, the SOPSs/SWSs are unable to respond proactively to adversary counterspace action. In fact, time could be misspent assessing other potential causes of the satellite anomaly. JSpOC ISRD has the responsibility for providing timely counterspace threat advisories and indications and warning (I&W), and it needs to determine and implement the best means by which to provide this information to wing intelligence units and SOPSs/SWSs.

---

[72] The "combined" in the name of CIC came about because CIC supported both U.S. Space Command and North American Aerospace Defense Command.

[73] We did not perform a manpower study. It is likely that the number could be much larger.

In the near term, JSpOC ISRD could forward this information by chat or email and update a website. It is our understanding that this is the process that JSpOC ISRD is currently undertaking, with information to be provided by Secret Internet Protocol Router Network (SIPRNET) and by Joint Worldwide Intelligence Communications System (JWICS) for those units that have access to top secret and sensitive compartmented information.[74] In the mid-term, AFSPC could define requirements for the JSpOC Mission System (JMS)[75] to be the mechanism for intelligence updates across the space enterprise. In the far term, this capability could then be phased into the JMS program.[76]

Another option to consider is to ensure that future satellites have sensors and instruments on board that can detect a range of threats and directly report to the SOPSs/SWSs so that they have real-time knowledge of possible adversary counterspace actions.[77] This option is not considered further because it is a materiel solution affecting the satellite and is, thus, outside the scope of this effort.

Space Knowledge of Intelligence Personnel

Third, most intelligence personnel assigned to space units, at the JSpOC level as well as at the squadron level, have limited knowledge about U.S. space systems or foreign counterspace capabilities because this is their first assignment to a space unit. This can impact resilience in that inexperienced intelligence personnel are not able to provide timely, actionable, and tailored intelligence to space operators in order for them to respond effectively to adversary counterspace actions. We recommend that the Air Force ensure that intelligence personnel assigned to JSpOC and to tactical units receive space training.

We understand that there is a space training pipeline backlog (discussed below in the "Finding: Personnel" section). This shortfall may be difficult to address in the near term, but certainly, in the mid-term, opportunities for formal space training for intelligence personnel assigned to operational space units should be increased. If possible, on subsequent assignment, it would be useful if more intelligence personnel were assigned to another space unit to retain their space knowledge, rather than having them return to the air community.[78]

---

[74] JSpOC ISRD, discussions regarding intelligence support to space operations centers, Vandenberg AFB, Calif., August 19, 2014.

[75] JMS uses an open systems architecture to provide applications, netcentric services and databases, and dedicated hardware to improve space situational awareness (SSA) and C2 of space forces. JFCC Space uses JMS to execute five lines of operation: space object identification, spectrum characterization, launch/reentry support, joint forces support, and support to contingency operations. JMS is upgraded using an incremental approach. Early versions of JMS have focused on SSA. Later versions will include more capabilities directly related to operational-level C2.

[76] We did not perform an assessment to determine the impact on the JMS program of incorporating intelligence updates as an application. This statement applies to other applications that we recommend below for incorporation into JMS.

[77] An example of such a capability is the Self-Awareness Space Situational Awareness (SASSA) program.

[78] During preparation of the final report, we received a comment regarding another mid-term option, and that was to investigate the possibility of creating a space specialty (i.e., an Air Force Specialty Code suffix, often referred to as a "shredout") for enlisted intelligence specialists. This would keep dedicated intelligence personnel in the space

Space Weather Effects

The fourth observation is related to space weather. We find that there is limited access at JSpOC and SOPSs/SWSs to timely and relevant space environment effects on satellite systems. During anomalies, JSpOC and SOPSs/SWSs will be unable to rapidly confirm or eliminate the space environment as the cause of an anomaly. We recommend that JSpOC Combat Operations Division (COD) determine and implement the best means by which to provide timely space weather effects advisories to the SOPSs/SWSs.

In the near term, this can be accomplished by forwarding information by chat or email and/or establishing a website with timely updates. In the mid-term, AFSPC should consider defining requirements for JMS to be the mechanism for providing space weather effects advisories (e.g., develop algorithms that automatically process relevant operational environmental data and turn into simple displays of effects on the user-defined operational picture). In the far term, these capabilities would then be phased into the JMS program.

Another option to consider is to ensure that future satellites have sensors and instruments on board that can detect space weather events and directly report them to the SOPSs/SWSs so that they have real-time knowledge of the event. Perhaps this capability could be integrated with the threat warning sensors discussed above. This option is not considered further because it is a materiel solution affecting the satellite and is, thus, outside the scope of this effort.[79]

Summary

The findings and implementation options discussed above are summarized in Tables 4.1 and 4.2, respectively. In the later cost section, the implementation options will be referred to by the mitigation number (in the left column of Table 4.2) and by the letter A, B, or C (in the other three columns of Table 4.2), which stand for near-term, mid-term, and far-term implementation options, respectively.

---

community. It would also make space intelligence their specialty and the skills they would be tested on for promotion. We did not evaluate this suggestion but thought it was worthwhile to include in our report.

[79] During preparation of the final report, the authors learned that the Air Force will integrate energetic charged particle sensors on all new satellite acquisitions (Deborah Lee James, Secretary of the Air Force, "Space Situational Awareness Energetic Charged Particle Monitoring Capability," memorandum, Washington, D.C., March 17, 2015).

Table 4.1. Findings: Information

Observation	Impact on Resilience	Mitigation
Space OB development resides at NASIC	Current space OB may not be readily available at the operational level (JSpOC)	Transfer space OB responsibility from NASIC to JSpOC ISRD
Limited to no access at SOPSs/SWSs to potential adversary counterspace force posture	SOPSs/SWSs are unable to respond proactively to adversary action; time could be misspent assessing other causes of a satellite anomaly	Determine and implement best means by which JSpOC ISRD can provide timely counterspace threat advisories and I&W to Wing Intelligence (INTEL) and SOPSs/SWSs
Intelligence officers assigned to space units may not be knowledgeable about U.S. space systems or about foreign counterspace capabilities	Inexperienced intelligence personnel are not able to provide timely, actionable, and tailored intelligence to space operators in order for them to respond effectively to adversary counterspace actions	Ensure that intelligence officers assigned to JSpOC and to lower-echelon space units receive space training
Limited access at JSpOC and SOPSs/SWSs to timely and relevant space environment effects	During anomalies, JSpOC and SOPSs/SWSs are unable to rapidly confirm or eliminate the space environment as the cause of an anomaly	Determine and implement best means by which JSpOC COD can provide timely space weather effects advisories to SOPSs/SWSs

Table 4.2. Implementation Options: Information

#	Mitigation	A: Near Term	B: Mid-Term	C: Far Term
1	Transfer space OB responsibility from NASIC to JSpOC ISRD	N/A	JSpOC ISRD dedicates staff to interface with NASIC staff (NASIC still to maintain space OB)	Establish a cadre (3 to 6) of government civilians with extensive knowledge of U.S. and foreign space capabilities at JSpOC ISRD to maintain space OB
2	Determine and implement best means for JSpOC ISRD to provide timely counterspace threat advisories and I&W to Wing INTEL and SOPSs	Forward information by chat or email and update website	Define requirements for JMS to be the mechanism for intelligence updates across the space enterprise	Phase capability into JMS program
3	Ensure intelligence officers assigned to JSpOC and to lower echelon space units receive space training	N/A	Increase opportunities for formal space training	N/A
4	Determine and implement best means for JSpOC COD to provide timely space weather effects advisories to SOPSs/SWSs	Forward information by chat or email and/or establish website with timely updates	Define requirements for JMS to be the mechanism for providing space weather effects advisories	Phase capability into JMS program

Findings: Organization and Tactics

JSpOC and the squadrons need to be correctly organized and possess the necessary tactics so that they can properly respond to an adversary's counterspace action. We have three observations that we believe have an impact on resilience.

Space Protection Lead

First, there is no position at JSpOC or at the SOPSs/SWSs assigned the duty to respond to adversary counterspace activity. We believe that this limits the various organizations' ability to provide timely, coordinated, and effective response at the operational and tactical levels. We recommend that JSpOC and the SOPSs/SWSs introduce a "space protection lead" who has the assigned responsibility to understand the threat environment, assess whether anomalies may be a result of intentional acts, and respond accordingly. Note that this would not absolve the other operators from also being on the lookout for anomalies and bringing them to the attention of the space protection lead and mission director. In addition, the space protection lead would ensure that current threat status was provided during crew shifts, just as the air community receives threat updates during the mission brief. The space protection lead would ensure that tactics and procedures were being developed as new threats to their systems appear. Further, they would interact with their counterparts in other space units to share and exchange lessons learned and best practices.

In the near term, this can be accomplished by assigning space protection as a secondary duty to an existing crew position. In the far term, as the threat level increases, it may be necessary to create a new crew position dedicated to this function at JSpOC and across the SOPSs/SWSs.[80]

Space Protection Tactics

Second, there is very limited development of tactics for responding to adversary counterspace action at the squadron level.[81] This clearly limits their ability to provide timely, effective, and coordinated response. We recommend that the squadrons establish a process for developing tactics for likely counterspace threats and make their development a priority. Today, squadron weapons and tactics elements are focused on developing tactics that support the force enhancement mission.

In the near term, SOPSs/SWSs weapons and tactics elements can develop job aids and procedures that address specific adversary counterspace threats; we find that some squadrons are doing so on their own initiative. These, of course, need to be documented and the crews trained

---

[80] During preparation of the final report, the authors learned that an operator position, a space protection duty officer, has been identified at JSpOC and that additional crew members are currently being trained in response to this need.

[81] While we focus here on preparing to respond to an adversary counterspace action, there are activities that can occur before any attack that contribute to resilience, to the extent that such activities are permitted or part of the ground element. They include avoiding patterns of behavior, utilizing redundancy, dispersing systems/capabilities, varying communication paths and patterns, employing tactical deception, and so forth in order to confuse an adversary. Developing tactics for such activities should be included; in this case, the "threat" may be, for example, the adversary's information-gathering capabilities.

on them through on-the-job training (OJT). While the weapons and tactics shops should take the lead, all space operators should be thinking about tactics development during training and throughout their operations career.

In the mid-term, the SOPSs/SWSs should submit these tactics to the formal TTP process for requisite testing and documentation (i.e., incorporation into an Air Force Tactics, Techniques, and Procedures [AFTTP] 3-1 document). Currently, Air Combat Command (ACC) owns this process, and it is not designed to accommodate the many and rapid changes that can occur at the various space operations centers. In the far term, it may be necessary to move the formal space TTP process under AFSPC to speed up what many see as a slow process. However, we note that ACC does have other options for more rapidly documenting TTP than the AFTTP 3-1 process (i.e., flash bulletins, tactics bulletins) that are not used by AFSPC. As a mid-term option, AFSPC should reevaluate its guidance document[82] and examine ways to streamline the formal TTP process and offer more options for rapid TTP documentation and dissemination.

Tactics-Sharing

Third, there is little to no information-sharing among operators of different U.S. space systems on potential responses to various adversary counterspace actions. This limits the ability of the SOPSs/SWSs to assess, plan, and execute a range of possible protection options that may be developed by other organizations. For example, the National Reconnaissance Office (NRO) may be developing tactics to protect its space assets that may be useful to the Air Force, but this information would be protected in highly classified compartments. Similarly, the Air Force is developing various offensive and defensive space control capabilities, again protected in highly classified programs, some of which may have a bearing on which threats the SOPSs/SWSs and NRO need to develop tactics against.[83] Information-sharing with civil space organizations is even more constrained because of their unclassified operations floor.

We recommend, in the near term, that the U.S. space community review and modify information sanitization procedures and protocols for granting temporary clearances to highly classified programs based on operational needs to increase information-sharing. Perhaps it can use the Coal Warfighter program, which provides key warfighters access to special capabilities, as a possible model.

Summary

The findings and implementation options discussed above are summarized in Tables 4.3 and 4.4, respectively.

---

[82] Specifically, AFSPC, "Tactics Development Program," AFSPC Instruction 10-260, Peterson AFB, Colo., November 29, 2011.

[83] During preparation of the final report, the authors learned that USSTRATCOM held its first Joint Space Doctrine and Tactics Forum in February 2015, at which operational leaders across the space enterprise discussed how to better prepare space forces to operate in an environment of increasing threats (ADM Cecil D. Haney, commander, U.S. Strategic Command, "Peter Huessy 'Space Power of the Warfighter' Breakfast Series," speech delivered at a breakfast seminar arranged by the Mitchell Institute, Washington, D.C., February 9, 2015.)

Table 4.3. Findings: Organization and Tactics

Observation	Impact on Resilience	Mitigation
No JSpOC or SOPS/SWS position assigned the duty to respond to adversary counterspace activity	Limits JSpOC and SOPS/SWS ability to provide timely, coordinated, and effective response at the operational and tactical levels	Introduce space protection lead at JSpOC and SOPSs/SWSs
Limited or no tactics for responding to adversary counterspace action	Limits SOPS/SWS ability to provide timely, effective, and coordinated response	Establish process for developing tactics for likely counterspace threats and make their development a priority
Lack of adequate information-sharing among operators of different U.S. space systems on potential responses to various adversary counterspace actions	Limits SOPS/SWS ability to assess, plan, and execute a range of possible protection options	Review/modify information sanitization procedures/protocols for granting temporary clearances based on operational needs (e.g., Coal Warfighter) to increase information-sharing

Table 4.4. Implementation Options: Organization and Tactics

#	Mitigation	A. Near Term	B. Mid-Term	C. Far Term
5	Introduce space protection lead at JSpOC and SOPSs/SWSs	Assign space protection as secondary duty to existing crew position	N/A	Create a new crew position across JSpOC and SOPSs/SWSs
6	Establish process for developing tactics for likely counterspace threats and make their development a priority	SOPS/SWS weapons and tactics develop job aids and procedures	SOPSs/SWSs submit tactics to formal TTP process for requisite testing and documentation AFSPC should reevaluate AFSPC Instruction 10-260 to streamline the formal TTP process and offer more options for rapid TTP documentation	Consider moving formal space TTP process under AFSPC to speed up process
7	Review/modify information sanitization procedures/protocols for granting temporary clearances based on operational needs (e.g., Coal Warfighter) to increase information-sharing	Review/modify information sanitization procedures/protocols for granting temporary clearances based on operational needs (e.g., Coal Warfighter) to increase information-sharing	N/A	N/A

Findings: Command and Control

The squadrons need to be able to contact their satellites in the event of adversary counterspace actions, and they need the authority to respond, perhaps in a very timely manner, to counter the threat. We have three observations that we believe have an impact on resilience.

Satellite C2 Contacts

While contacts are prioritized, the number and timing of satellite C2 accesses can be limited by the capabilities of the Air Force Satellite Control Network (AFSCN). For those constellations that rely on the AFSCN, space operators may not be able to readily contact their satellites for quick response to adversary counterspace actions.

One possible mitigation is for AFSPC to increase the number of available ground sites by leveraging allied and commercial capabilities. This may be of limited use today because these capabilities may not be compatible with existing U.S. space systems, but it could be a far-term option for future U.S. space assets if designed appropriately. It also would likely be a far-term option because of the need for extensive negotiations and the establishment of connectivity.

Another option is to consider adding satellite cross-linking to new systems. This option is not considered further because it is a materiel solution affecting the satellite and is, thus, outside the scope of this effort.

Responsibilities and Authorities

Authority for responding to adversary counterspace actions does not reside at SOPSs/SWSs; that authority resides with the commander, JFCC Space, and JSpOC. Depending on the threat, this could limit the ability of SOPSs/SWSs to provide a timely response. We recommend that JFCC Space review the chain of command and determine which command level should have responsibility and authority for various responses to adversary counterspace actions. We suggest that authority should reside at the lowest level possible to enable timely responses, but, clearly, this will be dependent on the level of conflict (e.g., theater conflict, strategic nuclear war). The tenet of centralized control and decentralized execution is considered crucial to the effective application of airpower. It may be time for the space community to relax its centralized control and centralized execution in certain situations, such as responding to adversary counterspace actions.

In the near term, the JSpOC Combat Plans Division (CPD) could develop courses of action (COAs) for likely adversary threats and establish rules of engagement (ROE) that authorize the lowest levels of command to provide more timely response. As noted above, the COAs and ROE will likely be dependent on the level of conflict. In the mid-term, AFSPC could define requirements for JMS to be the mechanism for enabling higher echelons to exercise "command by negation," in which lower levels are given authority to act but higher levels monitor the actions, and intervene when they deem necessary. Essentially, lower-level commanders report their intentions to act to a superior officer, noting that the action will be taken unless otherwise directed. In the far term, this capability can be phased into the JMS program.

Anomaly Resolution

It is commonly known that the current anomaly resolution process takes days and weeks (and frequently longer) and often comes to no resolution; this was confirmed during our discussions with the SOPSs/SWSs. This is not acceptable for time-critical events, such as adversary counterspace actions. This limits Air Force space authorities and SOPS/SWS ability to execute responses in a timely manner. We recommend that AFSPC ensure timely reporting of anomalies and develop a more timely anomaly resolution process. A possible first step is to review current space threats and force postures to rule out attack and then rule out space weather and orbital debris as possible sources of the anomaly. Then it can transition to a more standard, methodical process.

In the near term, JFCC Space should encourage SOPSs/SWSs to report anomalies in a timely manner to JSpOC; this would also enable JSpOC to learn whether there is a coordinated attack on U.S. space systems if multiple space operations centers report anomalies at the same time. In the mid-term, AFSPC should review the need for 24/7 availability of SMEs on duty at the squadrons to support a timely anomaly resolution process.[84] If necessary, additional SMEs would be added to the SOPSs/SWSs in the far term. In addition, in the mid-term, AFSPC should define requirements for improved diagnostics tools and then develop the tools in the far term.

Summary

The findings and implementation options discussed above are summarized in Tables 4.5 and 4.6, respectively.

Table 4.5. Findings: Command and Control

Observation	Impact on Resilience	Mitigation
While contacts are prioritized, number and timing of satellite C2 accesses can be limited by AFSCN capabilities	For those constellations that rely on AFSCN, space operators may not be able to readily contact their satellites for quick response to adversary counterspace actions	Investigate increasing the number of available ground sites by leveraging allied and commercial capabilities
Authority for responding to adversary counterspace actions does not reside at SOPSs/SWSs	Could limit SOPS/SWS ability to provide timely response	Review chain of command and determine which command level should have responsibility/authority for various responses to adversary counterspace actions (consider lowest levels)
Current anomaly resolution process is not timely enough for time-critical events	Limits Air Force space authorities and SOPS/SWS ability to execute responses in a timely manner	Ensure timely reporting of anomalies and develop a more timely anomaly resolution process

---

[84] Currently, 24/7 SME availability is met by "on-call" personnel.

32

Table 4.6. Implementation Options: Command and Control

#	Mitigation	A. Near Term	B. Mid-Term	C. Far Term
8	Investigate increasing the number of available ground sites by leveraging allied and commercial capabilities	N/A	N/A	Investigate increasing the number of available ground sites by leveraging allied and commercial capabilities
9	Review chain of command and determine which command level should have responsibility/authority for various responses to adversary counterspace actions	JSpOC CPD develops COAs for likely adversary threats and establishes ROE that authorize lowest levels of command to provide more timely response	Define requirements for JMS to be the mechanism for enabling higher echelons to exercise command by negation (monitoring of actions being taken by lower echelons)	Phase capability into JMS program
10	Ensure timely reporting of anomalies and develop a more timely anomaly resolution process (e.g., first rule out attack, space weather, orbital debris)	Encourage SOPSs/SWSs to report anomalies to JSpOC	Review need for 24/7 availability of SMEs Define requirements for improved diagnostics tools	Add additional SMEs to SOPSs/SWSs Develop improved diagnostic tools

Findings: Training

Space operators need to be trained to operate their systems in a contested environment. We have three observations that we believe have an impact on resilience.

Space Protection Training

We found that space operators are not formally trained to recognize and respond to adversary counterspace actions.[85] This is not surprising, given the lack of formally approved counterspace tactics and the lack of counterspace modules in training simulators and emulators. Training is focused on conducting the force enhancement missions, though undergraduate space training students are now being admonished to first rule out adversary effects, rather than assuming that an anomaly is due to a satellite system malfunction.[86] This lack of counterspace training can limit the SOPS/SWS ability to provide a timely and effective response. We recommend that AFSPC update the formal training process to include recognizing and responding to adversary counterspace actions.

---

[85] By formal training, we mean undergraduate space training (UST) and initial qualification training (IQT).

[86] 533 Training Squadron, discussions regarding counterspace training, Vanderberg AFB, Calif., April 3, 2014.

In the near term, the SOPSs/SWSs can establish OJT for the job aids and procedures that they have developed to address various counterspace threats; we find that some squadrons are doing so on their own initiative. In the mid-term, AFSPC should define requirements for adding counterspace training to the simulators and emulators now used for training.[87] However, to properly simulate the threat and its effects, it may be necessary to increase the fidelity of the simulators and emulators to ensure more realistic training; this would be part of the requirements definition process. Then, in the far term, AFSPC should upgrade the simulators and emulators or possibly acquire new ones with the necessary capabilities.

Exercises

Space operators do not get the same kind of training as does the air community on adversary threats, such as air defense. In particular, only a few exercises include space operators responding to adversary counterspace actions (although the air community receives training on the loss of space enablers, for example, caused by jamming satellite communications and the Global Positioning System).

In the mid-term, we recommend that, for current exercises (e.g., Red Flag, the Air Force's premier aerial combat training exercise held at Nellis AFB), AFSPC should provide sufficient funding, including travel funds to the exercise venue, for in-person participation by space operators from the force-enhancement community at the planning meetings as well as the actual exercises. This could increase the level of cross-domain interaction; in particular, the effect of enemy action on U.S. space systems would be more universally experienced and appreciated when it impacts a large force package. This would also require the construction of more robust, interconnected scenarios. As a second mid-term mitigation, AFSPC should develop new space exercises in which space operators respond to adversary counterspace actions and then ensure participation in these exercises with adequate funding.

Multiple Satellite C2 Systems

The Air Force uses many operational systems to command and control its satellites. This greatly limits the commander's flexibility in reassigning staff because certification is required per crew position and per space system. We recommend that AFSPC investigate changing the training regimen to increase the commander's flexibility in assigning staff to various crew positions or even different space systems. This could reduce the current operational limitation of having a finite number of crew members for each crew position.

In the mid-term, AFSPC could institute cross-training. There are two options: training for different crew positions on the same space system (i.e., on a system with which they are already familiar) or training for similar crew positions (e.g., space vehicle operator or payload operator) on different space systems. We note that this could be a costly option unless the training can be conducted at a local venue; with the 50th Space Wing and 460th Space Wing taking greater control of training, this option becomes more likely.

---

[87] More specifically, the mission system Training Planning Team should define requirements and review and update the Training System Requirements Analysis and the System Training Plan.

A more robust but far-term option is for AFSPC to develop a common ground system for C2 of all space systems. As operators progress in their career field, they would become more experienced with the common C2 system and would be more proficient at recognizing and diagnosing anomalies and other adverse events. AFSPC can review lessons learned from its Multi-Mission Space Operations Center (MMSOC) program. 1 SOPS uses MMSOC to provide C2 for a diverse set of satellites. AFSPC can also review Navy and commercial lessons learned on common C2 systems. The Navy has implemented the Common Ground Architecture for the satellites it commands and controls from Blossom Point. Similarly, the commercial satellite communications industry also flies a wide variety of satellite buses using a common satellite C2 system.[88] Along with the benefits, there are some risks with using a common C2 system. It could be a single point of failure from any "bugs" or other software problems and possibly from cyber attacks. However, over time, any problems will be found and corrected, and the C2 system can be isolated to protect from cyber attacks. If AFSPC decides to pursue a common C2 system (beyond MMSOC), it will need to weigh the benefits and risks.[89]

Summary

The findings and implementation options discussed above are summarized in Tables 4.7 and 4.8, respectively.

---

[88] Intelsat, discussions regarding commercial space operations and resilience, Washington, D.C., June 10, 2014.

[89] During preparation of the final report, the authors learned that AFSPC is studying a common C2 system. According to Gen John Hyten, commander of AFSPC, such a move would save money, improve cybersecurity, and make it far easier to train personnel (Andrea Shalal, "U.S. Air Force Moves Toward Common Satellite Control System," Reuters, April 16, 2015).

Table 4.7. Findings: Training

Observation	Impact on Resilience	Mitigation
Space operators are not formally trained to recognize and respond to adversary counterspace actions	Limits SOPS/SWS ability to provide timely and effective response	Update training process to include recognizing and responding to adversary counterspace actions
Only a few exercises include space operators responding to adversary counterspace actions	Space operators do not get the same kind of training as does the air community on adversary actions	For current exercises (e.g., Red Flag), provide sufficient funding, including travel funds, for participation by space operators from the force-enhancement community at planning meetings and actual exercises
		Develop new space exercises in which space operators respond to adversary counterspace actions and ensure participation with adequate funding
The Air Force uses many operational systems to command and control its satellites	Certification per crew position and space system greatly limits commander flexibility	Investigate changing the training regimen to increase commander flexibility in assigning staff to various crew positions or even different space systems

Table 4.8. Implementation Options: Training

#	Mitigation	A. Near Term	B. Mid-Term	C. Far Term
11	Update training process to include recognizing and responding to adversary counterspace actions	SOPSs/SWSs establish counterspace OJT for job aids and procedures	Define requirements for adding counterspace training to simulators and emulators	Upgrade simulators and emulators
12	For current exercises (e.g., Red Flag), provide sufficient funding, including travel funds, for participation by space operators from the force-enhancement community at planning meetings and actual exercises	N/A	For current exercises (e.g., Red Flag), provide sufficient funding, including travel funds, for participation by space operators from the force-enhancement community at planning meetings and actual exercises	N/A
13	Develop new space exercises in which space operators respond to adversary counterspace actions and ensure participation with adequate funding	N/A	Develop new space exercises in which space operators respond to adversary counterspace actions and ensure participation with adequate funding	N/A
14	Investigate changing the training regimen to increase commander flexibility in assigning staff to various crew positions or even different space systems	N/A	Institute cross-training	Develop a common ground system for C2 of space systems

Findings: Personnel

While we discuss personnel last, this is an important component of any operational concept. Space operators are the ones who are commanding and controlling space assets, and they need to do so in a space environment that is no longer benign. The Air Force needs the right personnel joining the career field, and it needs experienced personnel on console. We have three observations that we believe have an impact on resilience.

Initial Qualifications

Current qualifications to become space operators were developed when space was relatively benign. Thus, SOPS/SWS staff may not have the right technical or other qualifications or background to respond quickly, efficiently, and effectively to adversary counterspace actions.

In the near term, AFSPC should review and assess qualifications to become space operators, taking into account that space is now congested, contested, and competitive. Science, technology, engineering, and mathematics–cognizant (STEM-cognizant) degrees are now required of new space officers, and higher Armed Services Vocational Aptitude Battery (ASVAB) scores are required for enlisted personnel. While some experienced space operators argue that it is important to increase the average level of technical capability among those currently in the space career field, others argue that there is a need for a wider range of backgrounds, including the liberal arts. In particular, space operators may need to be adaptable, flexible, and creative when responding to an adversary counterspace threat, especially if there is no checklist for the particular situation. This topic requires further research that is beyond the scope of this project.

Career Progression

It appears to be increasingly difficult to maintain a cadre of experienced technical military operators. Unlike in the flying community, there are not a large number of assignments that both advance one's career as well as allow one to continue to gain experience in a specific weapon system. However, without an experienced technical space operations cadre, it may be difficult to develop and implement space protection measures.

In the mid-term, we recommend that AFSPC ensure that career progression and necessary technical skill acquisition and maintenance are effectively balanced and synchronized. This will require further research that is beyond the scope of this project. Perhaps the space career field needs to be split into generalists and specialists. Perhaps the highest-caliber operators with multiple tours can be placed in units with the most important missions, while the less critical or less vulnerable missions are used to grow the technical skill sets. Perhaps squadron crew manning needs modification. In the next observation, we discuss alternative manning that could potentially alleviate some of the tension between gaining and maintaining experience in a particular weapon system and career progression.

Trained Operators

Our final observation, a critical one, is that space operators generally spend less than one year on console once certified, although their tours are normally three years long. Because of a training pipeline backlog (i.e., the training program is not keeping up with the demand), newly assigned operators wait about a year before starting undergraduate space training at Vandenberg AFB, but the three-year clock starts ticking once they arrive at the squadron. As a result, crew members are unlikely to have the requisite experience and technical capability to quickly recognize and respond effectively to adversary counterspace actions.

For this observation, we list two mitigations. In the mid-term, it may be possible to extend active-duty tours to increase average experience level and technical capability on console; this will have the added benefit of reducing training demand but could have an impact on career progression.

In the far term, AFSPC may want to consider alternative manning (e.g., more Reserve, Guard, civilian, or contractor personnel) to increase the average experience level and technical capability of operators. Use of government civilians is not new to AFSPC; 22 SOPS operates the AFSCN, and the operators are all government civilians (except the commander), with long tenures at the unit. Contractors also provide subject matter expertise and engineering support at the SOPSs/SWSs. As discussed in Chapter 3, civil space agencies field experienced flight crews and encounter minimal turnover, with many operators generally having ten or more years of experience on console. Some civil agencies use government civilians and others a mix of government civilians and contractors. The commercial satellite communications companies also field experienced crews (again, generally having ten or more years of experience on console) and compensate them well to increase retention.[90] This also helps the companies contain the costs of satellite operations and training, which are expenses and not sources of revenue.

Summary

The findings and implementation options discussed above are summarized in Tables 4.9 and 4.10, respectively.

Table 4.9. Findings: Personnel

Observation	Impact on Resilience	Mitigation
Current qualifications to become space operators were developed when space was relatively benign (excluding space environment)	SOPS/SWS staff may not have the right technical or other qualifications or background to respond quickly, efficiently, and effectively to adversary counterspace action	Review and assess qualifications to become space operators, taking into account that space is now congested, contested, and competitive
It appears to be increasingly difficult to maintain a cadre of experienced technical military operators	Without an experienced technical space operations cadre, it may be difficult to develop and implement protection measures	Ensure that career progression and necessary technical skill acquisition and maintenance are effectively balanced and synchronized
Space operators spend less than one year on console once certified	Few crew members have the requisite experience to recognize and respond effectively to adversary counterspace action	Extend active-duty tours to increase average experience level and technical capability and to reduce training demand Consider alternative manning (e.g., more Reserve, Guard, civilian, or contractor personnel) to increase average experience level and technical capability and to reduce training demand

---

[90] Intelsat, 2014.

Table 4.10. Implementation Options: Personnel

#	Mitigation	A. Near Term	B. Mid-Term	C. Far Term
15	Review and reassess qualifications to become space operators, taking into account that space is now congested, contested, and competitive	Review and reassess qualifications to become space operators, taking into account that space is now congested, contested, and competitive	N/A	N/A
16	Ensure that career progression and necessary technical skill acquisition and maintenance are effectively synchronized	N/A	Ensure that career progression and necessary technical skill acquisition and maintenance are effectively synchronized	N/A
17	Extend active-duty tours to increase average experience level and technical capability and reduce training demand	N/A	Extend active-duty tours to increase average experience level and technical capability and reduce training demand	N/A
18	Consider alternative manning (e.g., Reserve, Guard, civilian, or contractor personnel) to increase average experience level and technical capability and reduce training demand	N/A	N/A	Consider alternative manning (e.g., Reserve, Guard, civilian, or contractor personnel) to increase average experience level and technical capability and reduce training demand

Cost of Implementation Options

A ROM cost for the implementation options discussed above is shown in Figure 4.1, in which the implementation option code (e.g., 17B) is used to represent each option.[91] We have characterized the costs as either low cost (less than $1 million), medium cost (between $1 million and $10 million), or high cost ($10 million or greater). Appendix A provides the estimating assumptions, data sources used, and the basis for the ROM costs. The options in the lower left of the table could be considered "low-hanging fruit"—i.e., those that are relatively easy to implement and are relatively low cost.

---

[91] The implementation option code refers to the mitigation number in the various implementation options tables (e.g., in the far left column of Table 4.10) and by the letter A, B, or C (noted in the other three columns of the tables), which stand for near-term, mid-term, and far-term implementation options, respectively.

**Figure 4.1. Implementation Options (for Each Mitigation):
Range from Low to High Cost Across Time Frames**

ROM Cost (FY14 $M)		Near-Term	Mid-Term	Far-Term
	High		17B	2C, 4C, 9C, 11C, 14C
	Medium	6A, 11A	2B, 4B, 6B, 9B, 13B, 14B	1C, 10C, 18C
	Low	2A, 4A, 5A, 7A, 9A, 10A, 15A	1B, 3B, 10B, 11B, 12B, 16B	5C, 6C, 8C

Timeframe

Low Cost	Up to $1M
Medium Cost	Greater than $1M and Up to $10M
High Cost	Greater than $10M

Near-term	< 1 yr
Mid-term	1 – 3 yr
Far-term	> 3 yr

We took a slightly different approach to arriving at our recommendations; that is, we did not focus solely on the lower left corner. We grouped and priced a set of near-term options and a set of far-term options that provide an end-to-end functional capability, based on the notional space protection operational concept discussed at the beginning of this chapter. Specifically, to accomplish the operational concept, we selected options from each of the elements discussed above: ensuring operators gain access to actionable information, assigning responsibility for taking action, developing counterspace tactics, defining responsive command and control procedures, and establishing relevant training and exercises.

In Figure 4.2, the specific near-term options are circled in red, and the far-term options are circled in purple. We also circled some mid-term options in purple that enable the far-term options. These became our detailed recommendations.

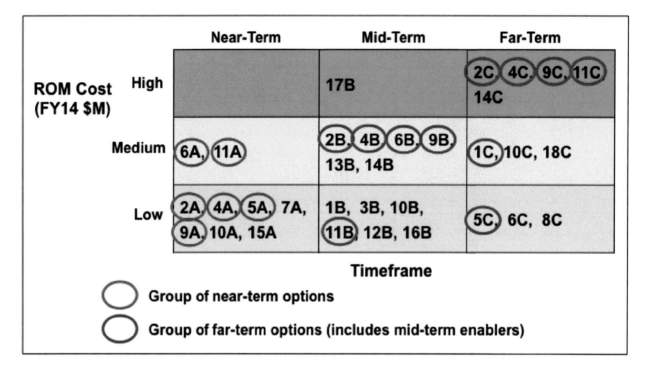

Figure 4.2. Group of Options That Provide an End-to-End Functional Capability

Detailed Recommendations

In Tables 4.11 and 4.12, we list the low-cost near-term implementation options and the more robust, more expensive, far-term implementation options, respectively. We also include the cost, with the range estimate providing a lower and an upper bound on where the most likely implementation cost falls. For the near-term options, we estimate the ROM NRE cost of implementation to be between $2.5 million and $3.6 million. Similarly, for the far-term options and the enabling mid-term options, we estimate the ROM NRE cost to be between $109 million and $166 million, and the ROM recurring (REC) cost to be between $4 million and $5.4 million per year. These two tables present our detailed recommendations.

Table 4.11. Low-Cost Near-Term Implementation Options

Mitigation	Near-Term Implementation Option	ROM Cost (Fiscal year [FY] 2014 dollars)
Determine and implement best means by which JSpOC ISRD can provide timely counterspace threat advisories and I&W to Wing INTEL and SOPSs/SWSs	Forward information by chat or email and update website (#2A)	$20,000–90,000
Determine and implement best means by which JSpOC COD can provide timely space weather effects advisories to SOPSs/SWSs	Forward information by chat or email and/or establish website with timely updates (#4A)	$20,000–90,000
Introduce space protection lead at SOPSs/SWSs and JSpOC[92]	Assign space protection as secondary duty to existing crew position (#5A)	$250,000–$380,000
Establish process for developing tactics for likely counterspace threats and make their development a priority	SOPS/SWS weapons and tactics develop job aids and procedures (#6A)	$1.0 million–$1.4 million
Review chain of command and determine which command level should have responsibility/authority for various responses to adversary counterspace actions	JSpOC CPD develops COAs for likely adversary threats and establishes ROE that authorize lowest levels of command to provide more timely response (#9A)	$180,000–$270,000
Update training process to include recognizing and responding to adversary counterspace actions	SOPSs/SWSs establish OJT for job aids and procedures (#11A)	$1.0 million–$1.4 million

---

[92] During preparation of the final report, the authors learned that an operator position, a space protection duty officer, has been identified at JSpOC and that additional crew members are currently being trained in response to this need.

Table 4.12. More-Robust Far-Term Implementation Options

Mitigation	Far-Term Implementation Option (Includes Mid-Term Enabler)	Nonrecurring ROM Cost (FY 2014 dollars)	Recurring ROM Cost (FY 2014 dollars)
Transfer space OB responsibility from NASIC to JSpOC ISRD	Establish a cadre of government civilians to maintain space OB (#1C)		$0.4 million–$0.7 million
Determine and implement best means by which JSpOC ISRD can provide timely counterspace threat advisories and I&W to Wing INTEL and SOPSs/SWSs	Define requirements for JMS to be the mechanism for intelligence updates across the space enterprise and phase capability into JMS program (#2B and #2C)	$13 million–$18 million	
Determine and implement best means by which JSpOC COD can provide timely space weather effects advisories to SOPSs/SWSs	Define requirements for JMS to be the mechanism for providing space weather effects advisories and phase capability into JMS program (#4B and #4C)	$46 million–$70 million	
Introduce space protection lead at SOPSs/SWSs and JSpOC[93]	Create a new crew position at SOPSs/SWSs and JSpOC (#5C)		$1.2 million–$1.7 million
Establish process for developing tactics for likely counterspace threats and make their development a priority	SOPSs/SWSs submit tactics to formal TTP process for requisite testing and documentation (#6B)		$2.0 million–$3.0 million
Review chain of command and determine which command level should have responsibility/authority for various responses to adversary counterspace actions	Define requirements for JMS to be the mechanism for enabling higher echelons to exercise command by negation and phase capability into JMS program (#9B and #9C)	$13 million–$18 million	
Update training process to include recognizing and responding to adversary counterspace actions	Define requirements for adding counterspace training to simulators and emulators and upgrade them (#11B and #11C)	$37 million–$60 million	

---

[93] During preparation of the final report, the authors learned that an operator position, a space protection duty officer, has been identified at JSpOC and that additional crew members are currently being trained in response to this need.

5. Resilience and a World with International and Commercial Partners

The 2011 *National Security Space Strategy* envisions a world in which the United States would "partner with responsible nations, international organizations, and commercial firms" and thus "ensure access to information and services from a more diverse set of systems—an advantage in a contested space environment."[94] The form of these agreements could resemble today's weather constellation, in which NOAA and the European Organisation for the Exploitation of Meteorological Satellites (EUMETSAT) collect operational environmental data from geostationary Earth orbits and polar low Earth orbits and share all of their data,[95] to the Air Force's agreement with SES Americom to manifest the Commercially Hosted Infrared Payload (CHIRP), a demonstration sensor employing wide field-of-view staring technology.[96] While all of the 18 suggested mitigations listed in the prior chapter would be affected in some manner by a new space architecture that includes international and commercial partners, four in particular require additional attention (2, 6, 9, and 10).

Information

Threat Advisories and Indications and Warning. We recommend that JSpOC provide timely threat advisories and indications and warnings to space operations centers. Much of this information would be classified, and most of what is classified is not releasable to foreign nationals, except perhaps to the Five Eyes community.[97] Clearly, this recommendation poses an implementation challenge in coalition and commercial environments. The U.S. government must weigh the benefit of releasing this data so that space operators can take appropriate actions in response to threats with the need to protect intelligence sources and methods.

Parallels to the space information-sharing quandary exist within the cyber domain. By Executive Order, the Department of Homeland Security, the Department of Justice, and the Director of National Intelligence are required to provide "timely production of unclassified reports of cyber threats to the U.S. homeland that identify a specific targeted entity."[98] Examination of how this is done, without compromising sources, methods, operations, and investigations, should yield insights for implementation at JSpOC.

---

[94] DoD and Office of the Director of National Intelligence, *National Security Space Strategy*, Unclassified Summary, Washington, D.C., January 2011, p. 8.

[95] NOAA, "NOAA, EUMETSAT Sign Long-Term Agreement for Weather, Climate Monitoring," news release, Silver Spring, Md., August 28, 2013.

[96] Air Force Space and Missile Systems Center, "Air Force Commercially Hosted Infrared Payload Mission Completed," news release, Los Angeles Air Force Base, Calif., December 6, 2013.

[97] "Five Eyes" refers to these nations: Australia, Canada, New Zealand, the United Kingdom, and the United States.

[98] White House, "Executive Order—Improving Critical Infrastructure Cybersecurity," Washington, D.C., February 12, 2013.

Another model for information sharing is USSTRATCOM's space situational awareness (SSA) sharing agreements. USSTRATCOM has signed agreements with EUMETSAT,[99] five nations (France, Japan, Australia, Italy, and Canada),[100] and approximately 35 commercial satellite owner/operators.[101] The agreements streamline the process for these partners to request specific SSA data that is gathered by JSpOC and that is important for launch support, satellite maneuver planning, support for on-orbit anomalies, electromagnetic interference reporting and investigation, satellite decommissioning activities, and on-orbit conjunction assessments.

Organization and Tactics

Counterspace Tactics. Developing and exercising counterspace tactics with allied nations is in the realm of the possible, but doing so with commercial partners could be problematic because of the chance the tactics may be compromised and then countered. This risk must be assessed before any disclosure. Moreover, working this closely with commercial partners may create unwelcome perceptions that commercial systems are legitimate infrastructure targets under the laws of war. However, commercial satellite communication providers who sell bandwidth to the U.S. military may already meet the criteria: "[I]f a civilian object makes an effective contribution to military action, and its destruction offers a definite military advantage, then it may be a legitimate target regardless of its civilian use."[102] In any event, commercial providers that contract with the U.S. military may find it to their advantage to learn how to defend themselves in cases of conflict, and the United States must decide to what extent it will support their efforts.

Command and Control

Courses of Action Development. COAs developed by JSpOC may involve leveraging commercial and coalition partners' space assets when U.S. military space assets are degraded. Formal agreements must be in place long before the start of a conflict and would need to be negotiated by USSTRATCOM or the Office of the Secretary of Defense, and possibly the Department of State. While commercial and coalition partners' space assets are not controlled by JSpOC, in the event of an attack, these partners could, for example, execute avoidance maneuvers while coordinating with JSpOC to enhance their survivability. Also, the development and exercise of such COAs with commercial and coalition partners would have to be assessed for disclosure risk.

---

[99] USSTRATCOM, "USSTRATCOM Enters into Space-Data Sharing Agreement with EUMETSAT," news release, Offutt Air Force Base, Neb., August 29, 2014.

[100] USSTRATCOM, "USSTRATCOM Signs Fifth Data Sharing Agreement," news release, Offutt Air Force Base, Neb., July 27, 2014.

[101] General C. Robert Kehler, commander, U.S. Strategic Command, remarks at the 29th National Space Symposium, Colorado Springs, Colo., April 16, 2013.

[102] Kristen Thomasen, "Air Power, Coercion, and Dual-Use Infrastructure: A Legal and Ethical Analysis," *International Affairs Review*, Vol. XVII, No. 2, Washington, D.C., Fall/Winter 2008.

Anomaly Reporting. Today, commercial and coalition partners do not report anomalies to JSpOC in a consistent manner. Centralized reporting may permit more timely recognition of coordinated attacks, including detection of less sophisticated adversary actions, such as intentional jamming. To provide incentive to commercial and coalition partners, JSpOC could offer value-added situational awareness information to its partners in return.

6. Recommendations

In the past, developers of NSS systems and architectures have not considered the full spectrum of possible counterspace actions that a potential adversary could undertake. This was due to a number of justifiable reasons, ranging from technology availability and military utilization to the geopolitical environment. Consequently, the mindset in the space community has been that "space is a sanctuary." This mindset resulted in space architectures and supporting ground infrastructures with limited capabilities to defend against adversary counterspace actions.

Despite that history, we found that AFSPC, at the headquarters level and at subordinate units, has begun addressing some of these limitations through improvements to both the space segment architecture—for example, in AoA activities—and the supporting ground infrastructure, which we designate as non-materiel (i.e., DOTMLPF-P). The non-materiel improvements being put in place (e.g., developing job aids for responding to some likely threats, adding resilience discussions to the training curriculum, and modifying personnel qualifications) address some of the shortfalls mentioned in this report. However, based on our research, there is more that can be done, and we summarize those recommendations below, beginning with two overarching recommendations.

Overarching Recommendations

Resilience as a Priority

One issue that was brought up by all the space operators with whom we met is the apparent lack of "demonstrated" priority to resilience by the leadership. Although they are aware that the AFSPC commander and other senior Air Force and DoD leaders from the space community discuss its importance,[103] the fact that resilience is a priority has not promulgated formally to space squadrons in the form of detailed implementation actions. Changing the prevalent mindset/culture within the rank and file of the space operator community will require that

- Space leadership define priorities and provide resources for non-materiel space resilience activities.

We expect that developing and implementing some of the recommendations provided in this report will help in both improving the resilience of the space enterprise as well as change the mindset of the involved personnel.

---

[103] The fact that U.S. satellites are at risk was made very clear to the nation on April 26, 2015, when CBS aired a segment called "The Battle Above" on *60 Minutes*, in which senior Air Force leadership described the contested space environment (CBS, "The Battle Above," *60 Minutes*, April 26, 2015).

Space Protection CONOPS

Because of the interdependence of the various DOTMLPF-P elements, making a few changes will not result in the desired improvement. We developed a set of implementation options to improve resilience based on a notional space protection operational concept—namely, enhancing the capability of space operators to respond, in a timely and effective manner, to adversary counterspace actions. To do so, operators need actionable information, appropriate organization and tactics, and dynamic C2, supported by appropriate tools and decision aids, relevant training and exercises, and qualified personnel brought into the career field. While this operational concept is a good starting point, we recommend that

- AFSPC develop a formal, end-to-end, space protection CONOPS that captures all the elements needed to improve resilience.

In addition, when developing the CONOPS, it may be time for the space community to relax its centralized control and centralized execution in certain situations, such as responding to adversary counterspace actions, and follow the tenet of centralized control and decentralized execution, which is considered crucial to the effective application of airpower.

Detailed Recommendations

As discussed in Chapter 4, we developed a detailed set of non-materiel implementation options to improve resilience based on our notional space protection operational concept. We grouped the recommendations for either near-term (less than one year) or far-term implementation (three to six years); they are summarized below.[104] First, we list the general mitigation to the identified shortfall, and then the specific implementation option for that timeframe.

Near-Term Recommendations

- Determine and implement the best means by which JSpOC ISRD can provide timely counterspace threat advisories and I&W to Wing INTEL and SOPSs/SWSs: Forward information by chat or email and update website.
- Determine and implement the best means by which JSpOC COD can provide timely space weather effects advisories to SOPSs/SWSs: Forward information by chat or email and/or establish website with timely updates.
- Introduce space protection lead at SOPSs/SWSs and JSpOC: Assign space protection as secondary duty to existing crew position.[105]

---

[104] Most of the mid-term implementation options were enablers of far-term options, and their costs were included in the far-term options.

[105] During preparation of the final report, the authors learned that an operator position, a space protection duty officer, has been identified at JSpOC and that additional crew members are currently being trained in response to this need.

- Establish process for developing tactics for likely counterspace threats and make their development a priority: SOPSs/SWSs weapons and tactics develop job aids and procedures.
- Review chain of command and determine which command level should have responsibility/authority for various responses to adversary counterspace actions: JSpOC CPD develops COAs for likely adversary threats and establishes ROE that authorize lowest levels of command to provide more timely response.
- Update training process to include recognizing and responding to adversary counterspace actions: SOPSs/SWSs establish OJT for job aids and procedures.

Far-Term Recommendations

- Transfer space OB responsibility from NASIC to JSpOC ISRD: Establish a cadre of government civilians to maintain space OB.
- Determine and implement best means by which JSpOC ISRD can provide timely counterspace threat advisories and I&W to Wing INTEL and SOPSs/SWSs: Define requirements for JMS to be the mechanism for intelligence updates across the space enterprise and phase capability into JMS program.
- Determine and implement best means by which JSpOC COD can provide timely space weather effects advisories to SOPSs/SWSs: Define requirements for JMS to be the mechanism for providing space weather effects advisories and phase capability into JMS program.
- Introduce space protection lead at SOPSs/SWSs and JSpOC: Create a new crew position.
- Establish process for developing tactics for likely counterspace threats and make their development a priority: SOPSs/SWSs submit tactics to formal TTP process for requisite testing and documentation.
- Review chain of command and determine which command level should have responsibility/authority for various responses to adversary counterspace actions: Define requirements for JMS to be the mechanism for enabling higher echelons to exercise command by negation and phase capability into JMS program.

ROM Costs

For the near-term options listed above, we estimate the ROM NRE cost of implementation to be between $2.5 million and $3.6 million. Similarly, for the far-term options listed above, we estimate the ROM NRE cost to be between $109 million and $166 million, and the ROM REC cost to be between $4 million and $5.4 million per year.

Appendix A: Space Resilience Cost Analysis

Cost Analysis Objective

This appendix describes the estimating assumptions, data sources used, and the basis for the ROM nonrecurring and recurring costs for implementing the set of near-term, mid-term, and far-term mitigation options for improving space resilience discussed in Chapter 4. Each ROM implementation cost is a range estimate that serves as an affordability metric to enable AFSPC decisionmakers to assess the relative magnitude of each implementation option compared to other options and, thus, help them better prioritize resources.

We have characterized the magnitude of the range estimates of each option as either a low cost (less than $1 million), a medium cost ($1 million or greater and less than $10 million), or a high cost ($10 million or greater). The range estimate provides a lower and an upper bound on where the most likely implementation cost falls. Each ROM range estimate also serves as an indication of the magnitude of the level of implementation cost uncertainty. Therefore, the ROM range estimates for each implementation option should not be used for budgetary purposes, but should serve to inform AFSPC decisionmakers in determining which options are potentially worth pursuing from an overall cost-benefit perspective.

Cost Estimating Overview

The estimating details provided here represent our initial basis for the ROM implementation options' costs, based on a combination of data inputs from

- *space operator SMEs* on the magnitude and/or complexity of the level of effort (LOE) required
- our collection of best available *relevant analogous data,* such as military and civilian salaries, JMS development program budgets and incremental plans, Standard Space Trainer (SST) technical and programmatic data, and other open source data
- our *research team's best engineering judgment,* especially for implementing options with varying levels of uncertainty.

We identify the costs as either being a one-time NRE estimate or annual recurring (REC) estimate. NRE estimates are associated with costs, for example, for developing new capabilities within the JMS program or space operator training aids and simulators. We associate NRE costs with mid- to far-term options in which we envision the implementation LOE and associated expenditures covering more than one year and, in some cases, up to five years. We also identify the costs estimated for a subset of implementation options in terms of the annual REC costs—for example, for covering the costs of increased cadre of qualified personnel required or for providing ongoing formal classroom space operations cross-training courses.

The next two sections provide a summary of the ROM range cost estimates for the set of near-term and far-term implementation options. The remainder of this appendix will be divided into four sections, with each section describing our ROM cost range estimating assumptions, the different data sources, and the basis of estimate approaches we used on a representative implementation option that impacts the majority of changes in one of the following four key areas:

- organizations and tactics
- personnel staffing
- space operations training
- space mission systems.

ROM Cost Summary of Near-Term Implementation Options

Table A.1 provides a summarized list of the ROM NRE cost range estimate results (all in FY 2014 constant dollars) we generated for implementing our list of six proposed options, each directly associated with a mitigation action (listed in the far left column). Note that all of the options are labeled with an "A" (e.g., option #2A) to indicate that we have considered each one as a "near-term" option, which we assumed would take less than one year for obtaining the potential funding[106] and approval to initiate. We estimated the cost for each of the four low-cost options and each of the two medium-cost options to take less than one year to implement. We estimated implementing all six near-term "low-hanging fruit" proposed options at a total ROM NRE annual cost ranging from $2.5 million to $3.6 million. Further details on the estimating assumptions, data sources used, and cost approach for option #9A, listed in Table A.1 in bold, are provided later in this appendix.

---

[106] Because of current budget constraints, decisionmakers may direct the action offices to use current funds and reprioritize their tasks, rather than request new funding. In the near term, this may be acceptable for many of the near-term implementation options.

Table A.1. ROM Cost of Near-Term Implementation Options

Mitigation	Near-Term Implementation Option	Nonrecurring ROM Cost (FY 2014 dollars)
Determine and implement the best means by which JSpOC ISRD can provide timely counterspace threat advisories and I&W to Wing INTEL and SOPSs/SWSs	Forward information by chat or email and update website (#2A)	$20,000–90,000
Determine and implement best means by which JSpOC COD can provide timely space weather effects advisories to SOPSs/SWSs	Forward information by chat or email and/or establish website with timely updates (#4A)	$20,000–90,000
Introduce space protection lead at SOPSs/SWSs and JSpOC	Assign space protection as secondary duty to existing crew position (#5A)	$250,000–$380,000
Establish process for developing tactics for likely counterspace threats and make their development a priority	SOPS/SWS weapons and tactics develop job aids and procedures (#6A)	$1.0 million–$1.4 million
Review chain of command and determine which command level should have responsibility/authority for various responses to adversary counterspace actions	**JSpOC CPD develops COAs for likely adversary threats and establishes ROE that authorize lowest levels of command to provide more timely response (#9A)**	$180,000–$270,000
Update training process to include recognizing and responding to adversary counterspace actions	SOPSs/SWSs establish OJT for job aids and procedures (#11A)	$1.0 million–$1.4 million

ROM Cost Summary of Far-Term Implementation Options

Table A.2 provides a summarized list of the ROM NRE range costs and REC annual range cost estimates (in FY 2014 constant dollars) associated with implementing seven options: one mid-term option (labeled with a "B") and six far-term options (labeled with a "C").

Note that ROM NRE range estimates for four out of the six far-term options also include our estimated cost for implementing four mid-term enabling options. Specifically, defining the requirements for improving JMS (listed as options #2B, #4B, and #9B) must be accomplished prior to phasing in these capabilities into the current JMS development program's plans (listed as options #2C, #4C, and #9C). Similarly, defining requirements for adding counterspace training to simulators/emulators (option #11B) must be accomplished prior to implementing the upgrade (option #11C).

We assessed the level of difficulty of the four enabling options as "mid-term," with a projected lead time for AFSPC and possibly higher chain of command coordination for the go-ahead approvals and funding to take between one and three years before initiating this set of options. For the set of options we proposed as "far-term," we assumed that it will require the Assistant Secretary of the Air Force (Acquisition) and potentially the Office of the Under Secretary of Defense for Acquisition, Technology, and Logistics higher chain of command coordination and approval for making program re-baseline changes as needed before awarding the contractor the contract and authority to proceed in implementing each option.

Table A.2. ROM Cost of Far-Term Implementation Options

Mitigation	Far-Term Implementation Option (includes mid-term enabler)	Nonrecurring ROM Cost (FY 2014 dollars)	Recurring ROM Cost (FY 2014 dollars)
Transfer space OB responsibility from NASIC to JSpOC ISRD	**Establish a cadre of government civilians to maintain space OB (#1C)**		$0.4 million–$0.7 million
Determine and implement best means by which JSpOC ISRD can provide timely counterspace threat advisories and I&W to Wing INTEL and SOPSs/SWSs	**Define requirements for JMS to be the mechanism for intelligence updates across the space enterprise and phase capability into JMS program (#2B and #2C)**	$13 million–$18 million	
Determine and implement best means by which JSpOC COD can provide timely space weather effects advisories to SOPSs/SWSs	Define requirements for JMS to be the mechanism for providing space weather effects advisories and phase capability into JMS program (#4B and #4C)	$46 million–$70 million	
Introduce space protection lead at SOPSs/SWSs and JSpOC	Create a new crew position at SOPSs/SWSs and JSpOC (#5C)		$1.2 million–$1.7 million
Establish process for developing tactics for likely counterspace threats and make their development a priority	SOPSs/SWSs submit tactics to formal TTP process for requisite testing and documentation (#6B)		$2.0 million–$3.0 million
Review chain of command and determine which command level should have responsibility/authority for various responses to adversary counterspace actions	Define requirements for JMS to be the mechanism for enabling higher echelons to exercise command by negation and phase capability into JMS program (#9B and #9C)	$13 million–$18 million	
Update training process to include recognizing and responding to adversary counterspace actions	**Define requirements for adding counterspace training to simulators/emulators and upgrade them (#11B and #11C)**	$37 million–$60 million	

As listed in Table A.2, we estimated the ROM NRE cost estimated for each of the four of the high-cost far-term and enabling mid-term set of proposed options to take between three and six years to implement. We estimated the total ROM NRE cost for implementing all four of these far-term proposed options concurrently to range from $109 million to $166 million, spanning this three- to six-year time frame. The total ROM REC estimate of between $4 million and $5.4 million per year represents the annual cost for implementing

- two far-term proposed options listed for adding and retaining qualified personnel within JSpOC (#1C), as well as the SOPSs and SWSs (#5C)

plus

- one mid-term option listed to cover the ongoing effort for the SOPSs and SWSs to submit tactics and to perform the requisite testing and documentation under a proposed formal TTP process (#6B).

Further details on the estimating assumptions, data sources used, and cost approach for three of the far-term options (see bolded items in Table A.2), covering a cadre of additional government civilian personnel costs (#1C), improving simulator training (#11B plus #11C), and implementing improved JMS capabilities (#2B plus #2C), are provided later in this appendix.

Representative Organization and Tactics–Related Near-Term Option

We proposed a near-term option for JSpOC CPD to develop COAs for likely adversary threats and establish ROE documentation for authorizing the lowest levels of command to provide timely responses depending on the threat and the specific COA.

We generated a ROM cost for this near-term effort based on the following assumptions. We assumed that CPD management would task its current qualified military personnel at the Air Force major and lieutenant colonel levels to develop the COAs and ROE documents. We assumed that it would take about a two to three full-time-equivalent (FTE) LOE staff spread over a six-month period for developing one to two COAs for countering three potential space threats. The assigned CPD staff would be tasked with generating one ROE documenting which command has the responsibility and authority to make decisions and how to respond to the threat and the selected COAs in a timely manner.

The cost for JSpOC CPD to implement this near-term organization and tactics-related option of generating a total of between three and six COAs, along with ROEs, is estimated at between 12 person-months (two FTE staff over six months) to 18 person-months (three FTE staff over six months) of additional effort. We estimated the military manpower ROM cost for this additional tasking effort to be between $180,000 and $270,000 (in FY 2014 dollars), based on the average Annual DoD Composite Rate[107] of 12 to 18 months of effort for a representative mix of CPD Air Force major (O-4) and lieutenant colonel (O-5) experienced staff in Air Force Special Code (AFSC) 13S3 qualified space operations positions[108] within JSpOC.

Even though the ROM cost represents the estimate for the additional effort required, we elected to use the annual DoD composite rates for Air Force officers over the lower military

---

[107] We based the implementation costs on the Annual DoD Composite Rate for Fiscal Year 2014 for AF military pay grade O-4s and O-5s cited in the "FY 2014 Military Personnel Composite Standard Pay and Reimbursement Rate" (John P. Roth, Office of the Under Secretary of Defense [Deputy Comptroller] Memorandum, "FY 2014 Department of Defense [DoD] Military Personnel Composite Standard Pay and Reimbursement Rates Memorandum, Attachment Tab K-5 Military Personnel Composite Standard Pay and Reimbursement Rates, Department of the Air Force for Fiscal Year 2014," May 9, 2013).

[108] The complete description of the Space Operations AFSC 13S3 Specialty Summary, Duties and Responsibilities, and Specialty Qualifications is documented in the "Air Force Officer Classification Directory (AFOCD)" (U.S. Air Force, "Air Force Officer Classification Directory [AFOCD]," April 30, 2013).

basic pay, since the additional assigned tasking for performing this option may, depending on the CPD overall workload, result in a JSpOC request for increased manpower levels and military personnel (MILPERS) budget, which includes the same cost elements used in the budget planning process. The annual DoD composite rates consist of average basis pay plus retired pay accrual, Medicare-eligible retiree health care accrual, basic allowance for housing, subsistence, incentive and special pay, permanent change of station expenses, and miscellaneous pay.

Representative Personnel Staffing Far-Term Option

In the far term, we proposed a personnel staffing–related option for transferring space OB responsibility from NASIC to JSpOC ISRD. We proposed in Table A.2 above, listed as Option #2C, establishing a cadre of between three to six government civilians capable of performing both intelligence functions and vulnerability assessments; and with sufficient familiarity along with military space operations for maintaining the space OB within ISRD.

We based the estimate on an assumed annual base civilian salary that includes an Air Force civilian personnel FY 2014 fringe benefits rate of 36.4 percent for labor costs incurred in support of reimbursable orders and another factor of 9.1 percent for the accrual of DoD civilian retirement, post-retirement health benefits, and post-retirement life insurance costs.[109] We applied these two factors to the FY 2014 General Schedule (GS) base pay for civilians for a cadre of a minimum of three and a maximum of six GS-10 through GS-12 personnel[110] with equivalent skills and experiences as an Air Force Space Operations Intelligence Officer (AFSC 14N3) with 12 months or more experience in INTEL ops functions.[111] The resulting ROM REC annual cost for adding the cadre of civilian personnel with JSpOC ISRD was estimated after applying both the 36.4-percent and 9.1-percent factors to the annual base pay at between $0.4 million and $0.7 million in FY 2014 constant dollars.

Representative Space Operations Training Mid- to Far-Term Options

We also proposed a space operations training–related set of options for improving JSpOC, SOPS, and SWS space operations staff skill levels for recognizing and responding to adversary threat through more timely engagement of counterspace actions. Even though there are near-term implementation options, such as OJT and use of training aids, we proposed, as listed in Table A.2 as mid-term option #11B, first defining space threat scenarios and specific training requirements for adding counterspace exercises as part of the detailed functions to be implemented in option #11C, either as

---

[109] The percentage rates are listed in John P. Roth, "Fiscal Year (FY) 2014 Department of Defense (DoD) Civilian Personnel Fringe Benefits Rates," OUSD Deputy Controller Memorandum, 2014.

[110] We used the average annual civilian base pay for GS-10 and GS-12 at Level 10 effective on the first day of the first applicable pay period beginning on or after January 1, 2014, as cited in White House, "Executive Order 13655—Adjustments of Certain Rates of Pay; General Schedule 1 (5 U.S.C. 5332(a)) for Civilians," Washington, D.C., December 23, 2013.

[111] U.S. Air Force, "Air Force Officer Classification Directory (AFOCD)," April 30, 2013, provides further details on the duties and responsibilities and specialty qualifications of intelligence officers (AFSC 14N3).

- an add-on simulator/emulator module within the SST

or

- open for bid to contractors as a stand-alone counterspace simulator to be used as part of the formal classroom space operations specialized training tool sets at AFSPC's Advanced Space Operations School (ASOpS).[112]

For mid-term option #11B, we assumed that the proposed effort would involve either a systems engineering technical assistance contractor or federally funded research and development center staff members working within AFSPC's space operations community to implement the requirements definition phase over a 12- to 18-month period. We scoped out the estimated LOE and number of FTE staff needed by assuming that the tasks involve, for example,

- collaborating with the JSpOC, SOPS, and SWS space operators, along with instructors at the Advanced Space Operations School
- gathering the key ROE and counterspace-related data needed
- defining the specific systems engineering top-down functional simulator adaptive training requirements and space operations user interface specifications at an adequate level of detail for implementing far-term option #11C of developing, testing, and delivering a counterspace simulator.

For far-term option #11C, we assumed that the proposed effort would be implemented either with

- the existing SST contractor, Sonalysts, Inc., for developing an add-on counterspace simulator module with expanded simulation features needed

or

- another contractor for developing a stand-alone simulator that specializes in implementing simulation-based space operator training and rehearsal capability products.

We estimated far-term option #11C as a three- to four-year effort, based on using two analogous cost data points from two Space and Missile Systems Center (SMC) previously awarded contracts with Sonalysts and making a best engineering judgment assessment of the effort needed for implementing option #11C compared to the complexity of these two previously awarded contracts. These two Sonalysts contracts cover the development of the

- Global Positioning System Next Generation Operation Control System Mission-Specific Plug-In training system modules, to operate within the SST environment, at a contract

---

[112] As one of the partners within the National Security Space Institute (NSSI), ASOpS expands space system understanding by providing world-class, in-depth instruction of space systems, capabilities, requirements, acquisition, strategies, and policies to support joint military operations and U.S. national security. ASOpS as cited at the NSSI website (NSSI, 2015), conducts (1) advanced training designed to train and educate space professionals in space warfighting tactics; (2) deployment training of space professionals on broad-based space applications, with an emphasis on theater integration; and (3) space operations training by introducing non-space professional students to space issues, policy, capabilities, limitations, and vulnerabilities.

value of $39.5 million (in then-year dollars), awarded on August 6, 2013, with an
expected completion date slightly over three years later on September 6, 2016[113]

- DoD Satellite Communications System mission-specific trainer, at a contract value of
$10.1 million (in then-year dollars), awarded on August 28, 2008.

The DSCS trainer contract was described as part of a Phase III overall effort that involved the
contractor procuring commercial off-the-shelf hardware and operating systems and developing
one common training system architecture capable of launching system-specific simulations
developed to execute space operations training for a number of different satellite systems.[114]

For the mid-term requirements definition phase (option #11B) plus the contractor
development far-term phase (option #11C) of producing, testing, and delivering a counterspace
scenario-based training simulator, we estimated a NRE range estimate of between $37 million
for an add-on module operating within the SST operating environment and $60 million for a
stand-alone training simulator (both in constant FY 2014 dollars).

Representative JMS Mid- to Far-Term Options

We also proposed a JSpOC capabilities improvement effort as part of the JMS program listed
in Table A.2 above as mid- to far-term options #2B and #2C: an automated approach for ISRD to
provide counterspace threat advisories and key I&W information to the Air Force Space Wings
INTEL, SOPS, and SWS operating units on a daily or more frequent basis.

Similar to the counterspace simulator set of options, we propose a requirements definition
phase (Option #2B) with the estimated LOE and staffing needed, for example, for the following:

- reviewing the relevant sources of the current intelligence information flow, volume of
data, and frequency of updates that exist currently and are being transmitted by JSpOC
through use of chat rooms or other classified forms of communication
- using this data-gathering assessment as the basis for setting the software functional
requirements and specification developed at the level of detail needed for phasing in the
incremental software development within the JMS program.

We estimated the LOE, acquisition timelines, and annual level of funding associated with
implementing incremental capabilities on JMS as the primary relevant analogous data source for
assessing, based on our best engineering judgment, the comparable major tasks and associated
ROM estimated costs for implementing both the mid-term and far-term options for the software
requirements definition, software development, coding, and testing necessary for delivering a
fully functional JSpOC database dissemination automated system. The automated system will
provide the key space counterspace threat advisory and I&W updates needed across the space
ground operations enterprise.

---

[113] The Sonalysts contract award notice on the AF SMC Range and Network Division cost-plus-fixed-fee contract
(FA8806-08-C-0001) was posted on August 6, 2013, at the DoD contracts website (U.S. Department of Defense,
"Contracts Press Operations, Notice No. 566-13," notice of Air Force contract award to Sonalysts, August 6, 2013).

[114] The Sonalysts contract award notice on the AF SMC contract (FA8806-08-C-0001) was posted on August 28,
2008, at the DoD Contracts website (U.S. Department of Defense, "Contracts Press Operations, Notice No. 724-08,"
notice of Air Force contract award to Sonalysts, August 28, 2008).

We assumed that the requirements definition phase would be covered as part of the JMS program's systems engineering, design, and development effort. As cited in the JMS FY 2015 program budget justification sheet details,[115] one of the analog JMS program data points we used for ROM sizing the time frames for the two options is that it takes, on average, approximately six months to develop the requirements for each JMS mission systems service pack application at a cost of approximately $6 million. It then takes another 12 to 24 months as a budget window for the mission applications software development phase, which includes software field-testing activities and independent verification and validation.

For both the mid-term JMS program requirements definition phase (option #2B) and the development far-term phase of producing, testing, and delivering this incremental capability to JSpOC ISRD (option #2C), we estimated an NRE ROM range estimate of between $13 million and $18 million (both in constant FY 2014 dollars) covering a span of between 20 and 30 months.

Summary Table of ROM Costs for All Implementation Options

Table A.3 summarizes our estimates of the ROM costs for all the near-term, mid-term, and far-term implementation options.

---

[115] Details on the JMS program annual funding, task descriptions, and development schedule timelines were extracted from U.S. Department of Defense, "Fiscal Year (FY) 2015 Budget Estimates, Air Force Justification Book, Volume 3B, Research, Development Test & Evaluation, Exhibit R-2," President's Budget (PB), submitted on March 2014.

Table A.3. Summary Table of ROM Costs for All Implementation Options

Option #	Near Term (A)	Mid-Term (B)		Far Term (C)	
	NRE Cost (FY 2014 millions of dollars)	NRE Cost (FY 2014 millions of dollars)	Annual REC Cost (FY 2014 millions of dollars)	NRE Cost (FY 2014 millions of dollars)	Annual REC Cost (FY 2014 millions of dollars)
1		$0.1			$0.4–$0.7
2	$0.02–$0.09	$2–$3		$11–$15	
3		$0.5–$0.9*			
4	$0.02–$0.09	$6–$10		$40–$60	
5	$0.25–$0.38				$1.2–$1.7
6	$1.0–$1.4		$2–$3		$0.1
7	$0.05				
8				$0.2	
9	$0.18–$0.27	$2–$3		$11–$15	
10	$0.05	$0.05–$0.10		$7–$10	
11	$1.0–$1.4	$0.05		$37–$60	
12		$0.05–$0.10		$7–$10	
13		$1.2–$1.8*			
14		$1.2–$1.8*		$30–$40	
15	$0.4				
16		$0.5–$0.9*			$15.5
17		$105–$175**			
18					$3–$7

* ROM estimates should be updated and sized based on a percentage of the annual space operations training budget and other factors.
** This item represents ROM MILPERS NRE manpower transition cost estimate impacts of extending military active-duty tours.

References

Preface

Department of Defense and Office of the Director of National Intelligence, *National Security Space Strategy*, Unclassified Summary, Washington, D.C., January 2011. As of February 2, 2016:
http://www.defense.gov/Portals/1/features/2011/0111_nsss/docs/ NationalSecuritySpaceStrategyUnclassifiedSummary_Jan2011.pdf

Summary

CBS, "The Battle Above," *60 Minutes*, April 26, 2015. As of February 2, 2016:
http://www.cbsnews.com/news/rare-look-at-space-command-satellite-defense-60-minutes/

Gruss, Mike, "Disaggregation Giving Way to Broader Space Protection Strategy," *Space News*, April 26, 2015. As of February 2, 2016:
http://spacenews.com/disaggregation-giving-way-to-broader-space-protection-strategy/

U.S. Department of Defense, "Space Policy," DoD Directive 3100.10, October 18, 2012. As of February 2, 2016:
http://www.dtic.mil/whs/directives/corres/pdf/310010p.pdf

Chapter 1

Defense Acquisition University, "DOTmLPF-P Analysis," ACQuipedia website, last updated on April 15, 2014. As of February 2, 2016:
https://dap.dau.mil/acquipedia/Pages/ ArticleDetails.aspx?aid=d11b6afa-a16e-43cc-b3bb-ff8c9eb3e6f2#anchorDef

Department of Defense and Office of the Director of National Intelligence, *National Security Space Strategy*," Unclassified Summary, Washington, D.C., January 2011. As of February 2, 2016:
http://www.defense.gov/Portals/1/features/2011/0111_nsss/docs/ NationalSecuritySpaceStrategyUnclassifiedSummary_Jan2011.pdf

Dreyer, Paul, Krista S. Langeland, David Manheim, Gary McLeod, and George J. Nacouzi, *RAPAPORT (Resilience Assessment Process and Portfolio Option Reporting Tool): Background and Method*, Santa Monica, Calif.: RAND Corporation, RR-1169-AF, 2016. As of April 2016:
http://www.rand.org/pubs/research_reports/RR1169.html

Gruss, Mike, "Disaggregation Giving Way to Broader Space Protection Strategy," *Space News*, April 26, 2015. As of February 2, 2016:
http://spacenews.com/disaggregation-giving-way-to-broader-space-protection-strategy/

Hura, Myron, Gary McLeod, and George J. Nacouzi, *Enhancing Space Resilience Through Non-Materiel Means: Appendix B—Missile Warning Mission Case Study*, Santa Monica, Calif.: RAND Corporation, 2016, not available to the general public.

Langeland, Krista S., David Manheim, Gary McLeod, and George J. Nacouzi, *How Civil Institutions Build Resilience: Organizational Practices Derived from Academic Literature and Case Studies*, Santa Monica, Calif.: RAND Corporation, RR-1246-AF, 2016. As of April 2016:
http://www.rand.org/pubs/research_reports/RR1246.html

Office of the Assistance Secretary of Defense for Homeland Defense and Global Security, "Space Domain Mission Assurance: A Resilience Taxonomy," white paper, Washington D.C., September 2015. As of February 2, 2016:
http://www.hostedpayloadalliance.org/getattachment/Resources/White-Papers/
Resilience-Taxonomy-White-Paper.pdf.aspx

U.S. Department of Defense, "Space Policy," DoD Directive 3100.10, October 18, 2012. As of February 2, 2016:
http://www.dtic.mil/whs/directives/corres/pdf/310010p.pdf

Chapter 2

Acton, J. M., and M. Hibbs, "Why Fukushima Was Preventable," Carnegie Endowment for International Peace, March 6, 2012.

Bernstein, A., D. Bienstock, D. Hay, M. Uzunoglu, and G. Zussman, "Power Grid Vulnerability to Geographically Correlated Failures—Analysis and Control Implications," *INFOCOM Proceedings*, IEEE, 2014.

Deloitte, "Supply Chain Resilience: A Risk Intelligent Approach to Managing Global Supply Chains," 2012. As of February 2, 2016:
http://www2.deloitte.com/content/dam/Deloitte/global/Documents/
Governance-Risk-Compliance/
dttl-grc-supplychainresilience-riskintelligentapproachtomanagingglobalsupplychains.pdf

Electric Consumer Research Council (ELCON), "The Economic Impacts of the August 2003 Blackout," February 2004.

Hirano, M., T. Yonomoto, M. Ishigaki, N. Watanabe, Y. Maruyama, Y. Sibamoto, et al., "Insights from Review and Analysis of the Fukushima Dai-Ichi Accident," *Journal of Nuclear Science and Technology*, Vol. 49, No. 1, January 2012.

Jennings, B. J., E. D. Vugrin, and D. K. Belasich, "Resilience Certification for Commercial Buildings: A Study of Stakeholder Perspective," *Environment Systems and Decisions,* No. 1779, 2013.

Kohn, Linda T., Janet M. Corrigan, and Molla S. Donaldson, eds., *To Err Is Human: Building a Safer Health System*, Washington, D.C.: National Academy Press, 1999.

LaPorte, Todd R., "High Reliability Organizations: Unlikely, Demanding, and at Risk," *Journal of Contingencies and Crisis Management,* Vol. 4, No. 2, June 1996.

Marchese, Kelly, and Jerry O'Dwyer, "From Risk to Resilience: Using Analytics and Visualization to Reduce Supply Chain Vulnerability," *Deloitte Review*, No. 14, 2014. As of February 2, 2016:
http://dupress.com/articles/dr14-risk-to-resilience/

Psychology Today, "Psych Basics: Resilience," website, New York, undated. As of February 2, 2016:
http://www.psychologytoday.com/basics/resilience

Roberts, K., "Some Characteristics of One Type of High Reliability Organization," *Organization Science,* Vol. 1, Issue 2, 1990.

Saenz, Maria Jesus, and Elena Revilla, "Creating More Resilient Supply Chains," MIT Sloan Management Review, Summer 2014. As of February 2, 2016:
http://sloanreview.mit.edu/article/creating-more-resilient-supply-chains/

SCRLC Maturity Model Team, "SCRLC Supply Chain Risk Management Maturity Model," interactive spreadsheet, April 2, 2013. As of February 2, 2016:
http://www.supplychainriskinsights.com/pdf/
scrlc_maturity_model_final_with_scoring_locked_2april2013.xlsx

Sheffi, Yossi, "Building A Resilient Supply Chain," *Harvard Business Review Supply Chain Strategy Newsletter,* October 2005.

Smith, J. P., and Gulfport CARRI team, "Organizational Resilience: Mississippi as a Case Study," Gulfport Resilience Essay of the Community and Regional Resilience Institute, March 2013.

Supply Chain Risk Leadership Council, "SCRLC Emerging Risks in the Supply Chain 2013," white paper, 2013. As of February 2, 2016:
http://www.scrlc.com/articles/Emerging_Risks_2013_feb_v10.pdf

TEPCO, "Fukushima Nuclear Accident Analysis Report," June 20, 2012.

Verni, C., "A Hospital System's Reponse to a Hurricane Offers Lessons, Including the Need for Mandatory Interfacility Drills," *Health Affairs,* Vol. 31, No. 8, 2012.

Weick, K., K. Sutcliffe, and D. Obstfeld, "Organizing for High Reliability: Processes of Collective Mindfulness," in R. S. Sutton and B. M. Shaw, eds., *Research in Organizational Behavior*, Vol. 1, Stanford, Calif.: Jai Press, 1999, pp. 81–123.

Weissman, J. S., C. L. Annas, A. M. Epstein, et al., "Error Reporting and Disclosure Systems: Views from Hospital Leaders," *JAMA: The Journal of the American Medical Association*, Vol. 293, 2005.

Wolf, Z. R., and R. G. Hughes, "Error Reporting and Disclosure," in R. G. Hughes, ed., *Patient Safety and Quality: An Evidence-Based Handbook for Nurses*, Rockville, Md.: Agency for Healthcare Research and Quality, 2008, Chapter 35.

Chapter 3

Air Force Research Laboratory, "Innovation: Threat Detection, Validation, and Mitigation Tool for Counterspace and Space Situational Awareness (SSA) Operations," SBIR Topic No. AF06-283, Wright-Patterson AFB, Oh., undated. As of February 2, 2016: http://www.afsbirsttr.com/Publications/Documents/ Innovation-092710-DF&NN-AF06-283.pdf

Diamond, Stephanie, and Anuj Marfatia, *Predictive Maintenance for Dummies*, Hoboken, N.J.: John Wiley and Sons, 2013.

NASA—*see* National Aeronautics and Space Administration.

National Aeronautics and Space Administration, discussions regarding NASA space operations and resilience with Goddard Space Flight Center staff, Greenbelt, Md., August 29, 2014a.

_____, discussions regarding NASA space operations and resilience with Johnson Space Center's Vehicle Integration Office staff, Houston, Tex., September 23, 2014b.

National Oceanic and Atmospheric Administration, discussions regarding NOAA space operations and resilience with Office of Satellite and Product Operations staff, Suitland, Md., September 4, 2014a.

_____, discussions regarding NOAA space operations and resilience with Office of Satellite and Product Operations staff, Suitland, Md., September 10, 2014b.

NOAA—*see* National Oceanic and Atmospheric Administration.

Tschan, C. R., and C. L. Bowman, *Development of the Defensive Counterspace Test Bed (DTB), Volume 1—Sensors and Detection*, TOR-2004(1187)-2, El Segundo, Calif.: Aerospace Corporation, 2004.

United States Geological Survey, "Landsat—A Global Land-Imaging Mission," USGS fact sheet 2012-3072, May 2013. As of February 2, 2016: http://pubs.usgs.gov/fs/2012/3072/fs2012-3072.pdf

_____, discussions regarding USGS space operations and resilience with Flight Systems staff, Greenbelt, Md., September 29, 2014.

USGS—*see* United States Geological Survey.

White House, *Assignment of Emergency Preparedness Responsibilities*, Executive Order 12656, Washington, D.C., November 18, 1988. As of February 2, 2016:
http://www.archives.gov/federal-register/codification/executive-order/12656.html

_____, *National Continuity Policy*, National Security and Homeland Security Presidential Directive, NSPD-51/HSPD-20, Washington, D.C., May 4, 2007. As of February 2, 2016:
http://policy.defense.gov/portals/11/Documents/hdasa/references/HSPD-20.pdf

_____, *National Space Policy of the United States of America*, Washington, D.C., June 28, 2010. As of February 2, 2016:
http://www.whitehouse.gov/sites/default/files/national_space_policy_6-28-10.pdf

Chapter 4

11 Space Warning Squadron, discussions regarding space operations and resilience, Schriever AFB, Colo., July 11, 2014.

11 SWS—*see* 11 Space Warning Squadron.

50 OG—*see* 50 Operations Group.

50 Operations Group, discussions regarding space operations and resilience with operations group staff and subordinate squadrons, Schriever AFB, Colo., February 19, 2014.

460 OG—*see* 460 Operations Group.

460 Operations Group, discussions regarding space operations and resilience with operations group staff and subordinate squadron, Buckley AFB, Colo., May 8, 2014.

533 Training Squadron, discussions regarding counterspace training, Air Education and Training Command, Vandenberg AFB, Calif., April 3, 2014.

533 TRS—*see* 533 Training Squadron.

614 Air and Space Operations Center, discussions with Operations Support Division staff regarding space operations and resilience at the operational level, Vandenberg AFB, Calif., August 19, 2014.

614 AOC—*see* 614 Air and Space Operations Center.

Air Force Space Command, "Tactics Development Program," AFSPC Instruction 10-260, Peterson AFB, Colo., November 29, 2011. As of February 2, 2016:
http://static.e-publishing.af.mil/production/1/afspc/publication/afspci10-260/afspci10-260.pdf

_____, discussions with Air, Space, and Cyberspace Directorate (A3) staff responsible for training, exercises, and tactics, Peterson AFB, Colo., various dates, 2014.

Haney, ADM Cecil D., commander, U.S. Strategic Command, "Peter Huessy 'Space Power of the Warfighter' Breakfast Series," speech delivered at a breakfast seminar arranged by the Mitchell Institute, Washington, D.C., February 9, 2015. As of February 2, 2016:

http://www.stratcom.mil/speeches/2015/126/
Peter_Huessy_Space_Power_for_the_Warfighter_Breakfast_Series/

Intelsat, discussions regarding commercial space operations and resilience, Washington, D.C., June 10, 2014.

James, Deborah Lee, Secretary of the Air Force, "Space Situational Awareness Energetic Charged Particle Monitoring Capability," memorandum, Washington, D.C., March 17, 2015.

Joint Staff, *Space Operations*, Joint Publication 3-14, Washington, D.C., May 29, 2013. As of February 2, 2016:
http://www.dtic.mil/doctrine/new_pubs/jp3_14.pdf

Joint Space Operations Center (JSpOC) Intelligence, Surveillance, and Reconnaissance Division (ISRD), discussions regarding intelligence support to space operations centers, Vandenberg AFB, Calif., August 19, 2014.

Shalal, Andrea, "U.S. Air Force Moves Toward Common Satellite Control System," Reuters, April 16, 2015. As of February 2, 2016:
http://www.reuters.com/article/2015/04/16/
usa-military-space-ground-idUSL2N0XD2HX20150416

United States Air Force, *Space Operations*, Doctrine Annex 3-14, Curtis E. LeMay Center for Doctrine Development and Education, Maxwell AFB, Ala., June 19, 2012. As of February 2, 2016:
https://doctrine.af.mil/download.jsp?filename=3-14-Annex-SPACE-OPS.pdf

Chapter 5

Air Force Space and Missile Systems Center, "Air Force Commercially Hosted Infrared Payload Mission Completed," news release, Los Angeles Air Force Base, Calif., December 6, 2013. As of February 2, 2016:
http://www.losangeles.af.mil/news/story.asp?id=123373357

Department of Defense and Office of the Director of National Intelligence, *National Security Space Strategy*, Unclassified Summary, Washington, D.C., January 2011. As of February 2, 2016:
http://www.defense.gov/Portals/1/features/2011/0111_nsss/docs/
NationalSecuritySpaceStrategyUnclassifiedSummary_Jan2011.pdf

Kehler, General C. Robert, commander, U.S. Strategic Command, remarks at the 29th National Space Symposium, Colorado Springs, Colo., April 16, 2013. As of February 2, 2016:
http://www.stratcom.mil/speeches/2013/91/29th_National_Space_Symposium/

National Oceanic and Atmospheric Administration, "NOAA, EUMETSAT Sign Long-Term Agreement for Weather, Climate Monitoring," news release, Silver Spring, Md., August 28, 2013. As of February 2, 2016:
http://www.noaanews.noaa.gov/stories2013/20130828_EUMETSAT.html

NOAA—*see* National Oceanic and Atmospheric Administration.

Thomasen, Kristen, "Air Power, Coercion, and Dual-Use Infrastructure: A Legal and Ethical Analysis," *International Affairs Review*, Volume XVII, No. 2, Washington, D.C., Fall/Winter 2008. As of February 2, 2016:
http://www.iar-gwu.org/node/40

U.S. Strategic Command, "USSTRATCOM Signs Fifth Data Sharing Agreement," news release, Offutt Air Force Base, Neb., July 27, 2014. As of February 2, 2016:
http://www.stratcom.mil/news/2014/464/
USSTRATCOM_Signs_Fifth_Data_Sharing_Agreement/

_____, "USSTRATCOM Enters into Space-Data Sharing Agreement with EUMETSAT," news release, Offutt Air Force Base, Neb., August 29, 2014. As of February 2, 2016:
http://www.stratcom.mil/news/2014/512/
USSTRATCOM_enters_into_Space-Data_Sharing_Agreement_with_EUMETSAT

White House, "Executive Order—Improving Critical Infrastructure Cybersecurity," Washington, D.C., February 12, 2013. As of February 2, 2016:
http://www.whitehouse.gov/the-press-office/2013/02/12/
executive-order-improving-critical-infrastructure-cybersecurity

Chapter 6

CBS, "The Battle Above," *60 Minutes*, April 26, 2015. As of February 2, 2016:
http://www.cbsnews.com/news/rare-look-at-space-command-satellite-defense-60-minutes/

Appendix A

National Security Space Institute, home page, 2015. As of February 2, 2016:
https://www2.peterson.af.mil/nssi/public/

NSSI—*see* National Security Space Institute.

Roth, John P., Office of the Under Secretary of Defense (Deputy Comptroller) Memorandum, "FY 2014 Department of Defense (DoD) Military Personnel Composite Standard Pay and Reimbursement Rates Memorandum, Attachment Tab K-5 Military Personnel Composite Standard Pay and Reimbursement Rates, Department of the Air Force for Fiscal Year 2014," May 9, 2013.

_____, "Fiscal Year (FY) 2014 Department of Defense (DoD) Civilian Personnel Fringe Benefits Rates," OUSD Deputy Controller Memorandum, 2014. As of February 2, 2016:
http://comptroller.defense.gov/Portals/45/documents/rates/fy2014/2014_d.pdf

U.S. Air Force, "Air Force Officer Classification Directory (AFOCD)," April 30, 2013.

U.S. Department of Defense, "Contracts Press Operations, Notice No. 724-08," notice of Air Force contract award to Sonalysts, August 28, 2008. As of February 2, 2016: http://archive.defense.gov/contracts/Contract.aspx?ContractID=3851

———, "Contracts Press Operations, Notice No. 566-13," notice of Air Force contract award to Sonalysts, August 6, 2013. As of February 2, 2016: http://archive.defense.gov/Contracts/Contract.aspx?ContractID=5103

———, "Fiscal Year (FY) 2015 Budget Estimates, Air Force Justification Book, Volume 3B, Research, Development Test & Evaluation, Exhibit R-2," President's Budget (PB) submitted on March 2014.

White House, "Executive Order 13655—Adjustments of Certain Rates of Pay; General Schedule 1 (5 U.S.C. 5332(a)) for Civilians," Washington, D.C., December 23, 2013.